International Pesticide Product Registration Requirements

ACS SYMPOSIUM SERIES **724**

International Pesticide Product Registration Requirements

The Road to Harmonization

Willa Garner, EDITOR
GARNDAL Associates, Inc.

Patricia Royal, EDITOR
Quality Systems Consultants

Francisca Liem, EDITOR
Environmental Protection Agency

American Chemical Society, Washington, DC

Library of Congress Cataloging-in-Publication Data

International pesticide product registration requirements : the road to harmonization / Willa Garner, editor, Patricia Royal, editor, Francisca Liem, editor.

p. cm.—(ACS symposium series , ISSN 0097–6156 ; 724)

Includes bibliographical references and index.

ISBN 0–8412–3599–6 (alk. paper)

1. Pesticides—Toxicology. 2. Pesticides—Environmental aspects. Pesticides—Law and legislation.

I. Garner, Willa Y., 1936– . II. Royal, Patricia. III. Liem, Francisca. IV. Series.

RA1270.P4I59 1999
615.9′.51—dc21 99–24130
CIP

The paper used in this publication meets the minimum requirements of American National Standard for Information Sciences—Permanence of Paper for Printed Library Materials, ANSI Z39.48–1984.

Distributed by Oxford University Press

PRINTED IN THE UNITED STATES OF AMERICA

Advisory Board

ACS Symposium Series

Foreword

THE ACS SYMPOSIUM SERIES was first published in 1974 to provide a mechanism for publishing symposia quickly in book form. The purpose of the series is to publish timely, comprehensive books developed from ACS sponsored symposia based on current scientific research. Occasionally, books are developed from symposia sponsored by other organizations when the topic is of keen interest to the chemistry audience.

Before agreeing to publish a book, the proposed table of contents is reviewed for appropriate and comprehensive coverage and for interest to the audience. Some papers may be excluded in order to better focus the book; others may be added to provide comprehensiveness. When appropriate, overview or introductory chapters are added. Drafts of chapters are peer-reviewed prior to final acceptance or rejection, and manuscripts are prepared in camera-ready format.

As a rule, only original research papers and original review papers are included in the volumes. Verbatim reproductions of previously published papers are not accepted.

ACS BOOKS DEPARTMENT

Contents

Preface

The symposium on which this book is based helped to build an understanding of the different perspectives in international pesticide product registration procedures. It focused discussion on harmonizing data quality, technical procedures, and the regulatory process with the goal of minimizing barriers to universal data acceptance and usage. So, as we traveled together down this road toward mutual acceptability, we found the similarities to be numerous and the differences to be relatively few.

Our intent and focus was manifold: First, to raise the level of understanding in the different registration procedures ongoing today; second, to discuss the international pesticide product registration process; and third, to provide a forum to synchronize technical and regulatory requirements for producing universal data quality standards. Our speakers came from India, Australia, Japan, the Netherlands, Germany, France, England, Switzerland, Brazil, Argentina, Canada, Mexico, and the United States. They included government officials and representatives from industry, such as sponsors, research contract laboratories, and field consultants. They presented a comprehensive perspective of where we are today with our current policies and the direction to move in the future to promote universal data reciprocity. Topics included technical study guidelines, submission requirements, pesticide policy assessment, and international standards for good laboratory practices, including quality assurance and ISO accreditation of laboratories.

As coorganizers of the symposium (along with Richard Turle, the Environmental Technology Center, Environment Canada, and Amada Velez, Secretaria de Agricultura Ganaderia y Desarrolo Rural, Direccion General de Sanidad Vegetal, Direccion de Servicios y Apoyo Tecnico, Mexico) and the editors of this volume, we thank contributors and reviewers, whose expertise and generosity with their time will make this book a valuable reference for those working in international pesitcide registration programs. As editors of this publication, we are indeed grateful. We give special thanks to the editorial staff of ACS books for their support in this effort.

We also express our appreciation to American Agricultural Services, Inc., the American Crop Protection Association, American Cyanamid, Centre Analytical Laboratories, Inc., Dow Agrosciences, DuPont Agricultural Products, Fifth Chemical Congress of North America, Lonza Inc. Research and Development, and Novartis for their abiding interest and financial support and to the Division of Agrochemicals of the American Chemical Society for sponsoring this forum.

Willa Garner
GARNDAL Associates, Inc.
17485 Sierra Way
Monument, CO 80132

Patricia Royal
Quality Systems Consutants
80 Main Street
Plympton, MA 02367

Francisca Liem
Laboratory Data Integrity Branch
Office of Compliance, Environmental Protection Agency
401 M Street, SW
Washington, DC 20460

Chapter 1

Overview

Richard Turle

Environmental Technology Centre, Environment Canada, Ottawa K1A 0H3, Canada

The quest for quality of analytical measurements, just to take one field of scientific endeavor used in the development of pesticides, has long been the aim of analytical chemistry. Irrespective of the nature of the material being analyzed, analytical chemists have tried to produce quality results. Indeed, to ensure quality results, bodies such as The Association of Official Analytical Chemists were established in the last century to apply standardized test methods in the hope that accuracy would follow. Eventually, the terms accuracy and precision when applied to a numerical value obtained from analysis defined the quality of the result. Modern instrumentation and the application of computers and statistical software have made the production of apparently accurate and precise results a seemingly easy thing to do. Yet today we are still on a continued quest for improving the quality of analytical results and associated data. Certainly, in the area of pesticide registration, the quest continues unabated, and it is not just the analytical chemist who is concerned. It is the toxicologist, the field and animal scientist, the regulator, and the environmentalist who also are concerned with the quality of the data. Why?

The answer lies in the need for the public to be assured that the products that they use are safe. Safe is a big word in the context of pesticide registration. Safe can be defined as safe in terms of specificity of effect, safe against harmful effects for the transporter, applicator or bystander; safe against health effects for the consumer of the target organism , whether it be a direct or incidental application; and finally, safe for the environment in that a healthy ecosystem will remain after application and the pest has been eliminated. Given that pesticides, unlike pharmaceutical products, are designed to kill something, this indeed a tall order. The responsibility for ensuring a pesticide is safe falls in most countries to a form of regulating body. Such regulating bodies have the responsibility to make a judgment on the validity of scientific data generated by pesticide producers. In Canada, this is now the Pest Management Regulatory Agency that reports to the federal Minister of Health. In the USA, it is the Environmental Protection Agency which administers the Federal Insecticide, Fungicide and Rodenticide Act. In Mexico, the responsibility falls to the Departments of Agriculture and Health. Other OECD member countries have similar agencies for pesticide registration. Just to complicate a technically complex problem, pesticides are often developed in one country, manufactured in a second, and then shipped to a third before final use. Obviously, any system to ensure safety of pesticides must recognize these realities of a global economy.

Is the public right to be concerned with safety? Undoubtedly, yes! Ever since Rachel Carson's far-seeing book, "The Silent Spring" (*1*), described the effects of organochlorine pesticides, there has been an increasing awareness that it has not been acceptable for a pesticide manufacturer, user or regulating body to say a pesticide is safe without a considerable quantity of data being generated to prove it. Producing such data is expensive and time consuming. The registration process is costly, and without registration, a pesticide is without commercial value or benefit to the farmer. Given these economic and sometimes conflicting pressures, it is essential that systems exist to ensure that all of the scientific data used for registration, whether produced in the laboratory or in the field, have been acquired under known and verifiable conditions with every aspect documented. Such systems must be inherently strong enough to ensure that errors, incompetence or outright fraud are discovered and corrected before registration is allowed. For such reasons, scientific data gathered for the purposes of pesticide registration must be of known quality and will have had to have been gathered under a quality management system.

Generally, two systems have been developed to ensure that scientific measurements and test results are accurate. These two systems, while not incompatible in a single laboratory, have different aims and are driven by different needs. The most common one in use today is based on ISO Guide 25 (*2*). This is a technical standard for a quality system which is used by many countries to allow laboratories to gain accreditation. This assures both clients of laboratories and regulating bodies, which may receive their data, that the laboratory has a quality system in place, with written test methods and standard operating procedures, which is managed by a quality assurance officer.

The certificate of accreditation, given by a national accrediting organization, is given only after a site inspection conducted by qualified auditors. Inherently, such accreditation is most suitable where statistical quality assurance can be applied. In other words, it best applies to situations where there are many samples for each test. It is the most common quality system used by laboratories which have to confirm that products in trade meet established standards. Accredited laboratories are used by regulatory bodies in many countries, post registration, to ensure that food does not contain unacceptable levels of pesticides. No attempt is made in such accredited laboratories to ensure that **all** data can be subject to a later audit or reconstruction.

Good Laboratory Practice Standards (GLPS) is another quality management system that has been applied to both laboratory and field testing for pesticide registration, as well as to the pharmaceuticals, chemical substances, and food additives registration processes. The driving force behind GLPS as a quality system is to ensure the regulators that all of the data produced are accurate and that, if required, it is possible to reconstruct the results from the raw data. Inherently, this system works best when applied to a situation of having many tests applied to one substance. In many ways, and indeed, in some laboratories, the two systems require the same kind of information: test method procedures, sample handing procedures, etc. Another unique feature for GLP is that a study director is required. It is this person's responsibility to ensure studies are conducted to the GLP requirements for the purpose intended. It is a mandatory requirements that there is a quality assurance officer, and that there is sufficient control of documentation to ensure that any study can be fully reconstructed. Both the study director and the QA officer mutually rely on each other to ensure that only effective data are produced. The scope of the required documentation is far beyond that required by ISO Guide 25. For example, if the test report indicates that pesticide residue studies were conducted on 1000 mice, there should be records to indicate that at least 1000 mice were purchased and that their taxonomy was confirmed by a veterinarian.

How do these two quality systems relate to the demands of pesticide product registration requirements? Most countries requiring registration of pesticides require data submitted to have been prepared according to one of the forms of GLP in use today. These are the OECD (*3*) and American forms of GLP, as mandated by the USFDA and the USEPA. Both have the same intent and eventually may be harmonized into one system. Increasingly, developing countries are also making GLP a requirement for pesticide registration. In some countries, the use of ISO Guide 25 in laboratories is being recognized as an acceptable form of quality system for post registration studies in assessing actual application rates, application concentrations, and food chain residues.

The codification of quality systems, such as GLP or ISO Guide 25, does not remain static. Issues, such as verifying the accuracy of computer systems in data capture and data interpretation and retrieval, are in a state of flux. The application of GLP to field studies, where there is far less control over environmental conditions than in a laboratory, has been an area of debate among the GLP practitioners and pesticide companies and regulators. The GLP issues in the field relate to recording application conditions, weather conditions and other factors that might affect the results. The cost of pesticide regulation and the cost of producing information to obtain a registration increasingly has led to recognition that there needs to be some form of international harmonization between the various registration systems. As a direct consequence, there is a need to ensure that the GLP systems also are harmonized among themselves so there is complete recognition internationally for any one country's test data. Further, there is a debate among quality assurance practitioners that there should be an examination of quality systems, such as ISO Guide 25 and GLP, to identify commonalities so that only one accreditation is required for laboratories working under both systems. The obstacles to achieve this are formidable; however, such achievement would result in efficiency within laboratories accredited for both systems.

In this symposium volume, various details of the quality systems are presented as they are applied by different countries to these issues, as well as to other aspects of the increasing number of requirements for pesticide registration.

Literature Cited

1. Rachel Carson, *Silent Spring*, Houghton Mifflen, Boston, Mass., 1962.
2. ISO Guide 25: 1990, *General Requirements for the Competence of Calibration and Testing Laboratories.*
3. The OECD Principles of Good Laboratory Practices, Environment Monograph No. 45; Organization for Economic Cooperation and Development, Paris, 1992.

Chapter 2

International Trade Agreements and Their Impact on Health and Environmental Standards and Quality Assurance of Data

Irving L. Fuller

**U.S. Environmental Protection Agency,
401 M Street, S.W., Washington, DC 20460**

The countries that are parties to the World Trade Organization and its related agreements (WTO) have incurred certain international rights and obligations in the areas of health and environmental standards and quality assurance of data. Similar provisions exist in the North American Free Trade Agreement (NAFTA) which obligate Mexico, Canada and the United States of America. The rights and obligations related to the agreements involving Sanitary and Phytosanitary Measures are inter-alia relevant to international pesticide requirements. These trade agreements also impact conformity assessment procedures such as the OECD Good Laboratory Practices (GLPs) and the recent requirements regarding the use of voluntary consensus standards by U.S. government agencies under the National Technology Transfer and Advancement Act of 1995, as well as the relationship of voluntary consensus standards to the provisions in the trade agreements regarding the use of international standards.

The Major Factors

The impressive growth of world trade and globalization of industries has been triggered by many factors. Significant influences include: the end of the cold war, the collapse of the Soviet Union, the emergence and rapid transition of newly independent states, the establishment and exponential growth of the European Union, and the establishment of the new World Trade Organization (WTO) and one of the WTO consistent regional agreements - - the North American Free Trade Agreement (NAFTA). In addition, modern manufacturing, packaging, storage, preservation, and transport now make possible worldwide and regional movement of many commodities so perishable that until recently only locally grown crops could be stocked or sold.

Increased trade requires greater cooperation among the trading partners to minimize any potential health, safety, or environmental problems. This is especially true for trade with countries that may lack adequate regulatory infrastructures to register pesticides or to ascertain and control levels of pesticide residues in food.

All these factors combine to create an unprecedented demand for agreements among nations regarding aspects of regulatory procedure which impact directly or indirectly the movement of goods in commerce. Such issues as local, state and national approval processes for agricultural and industrial products are the subject of intense interest. These include all of the steps leading up to approval and may, if countries agree, include the final regulatory approval for goods to move in commerce. Industry understandably wants to move its products as rapidly as possible from the factory or farm to users around the globe. At the same time nations and sub-national jurisdictions wish to preserve their rights to take reasonable steps to properly protect health, safety, and the environment in their territories. Environmental groups express grave concern about damage to ecosystems and natural resources caused by invasions of harmful xenobiotic species. The President's new initiative to protect against illnesses caused by imported fruits and vegetables is in response to several incidents of illness related to contaminants on imported fruits and vegetables.

How can all these disparate and seeming to be conflicting demands and hopes be met? This calls for the wisdom of Solomon, and yet bureaucracy is rarely seen to possess such intelligence. The now retired but much admired humorist Gary Lawson, creator of a comic strip called "The Far Side," captured a similar situation with a cartoon showing a large group of dinosaurs being lectured by one of their peers about the grave challenges facing their species including the danger of asteroids, climate change, the rise of the mammals, etc. The reptilian lecturer concluded - "and to deal with these problems we have a brain the size of a walnut." Hopefully the bureaucratic brain, reptilian though it may appear to many, is sufficiently larger to allow us to develop slightly more strategy than that evidenced by the dinosaurs.

The Negotiation of Trade Agreements that Attempt to Balance Health, Environment and Trade Concerns

The World Trade Organization (WTO) was a principal product of the Uruguay Round of Trade Negotiations (*1, 2*) which created for the first time a permanent secretariat to assist nations in carrying out their rights and obligations under the WTO agreements.

The WTO agreements consist of the GATT '94, which contains the latest and last amendments to the General Agreement on Tariffs and Trade (GATT) and related agreements. This remarkable instrument was created in 1947 in the aftermath of World War II in an attempt to promote world trade. Part of the motivation was the concern that significant barriers to trade had contributed to the causes of World War II. One may discern from The GATT's title that the primary aim during most of its evolution was on the reduction of tariffs among its contracting parties. As it succeeded in this goal, the focus began to shift toward the reduction of non-tariff trade barriers. The Tokyo Round of trade negotiations in 1979 saw the emergence of a Voluntary Standards Code which attempted to ensure the legitimate rights of nations to impose regulations, laws, etc., to protect human health and the environment while beginning a process for Contracting Parties of the GATT, which chose to join the Code to reduce and, where possible, eliminate restrictions on trade from so called "non-tariff" barriers.

It was at this point in the evolution of the GATT that agencies such as EPA and FDA began to be involved in the negotiation of trade agreements since environmental, health and safety regulations can pose barriers, albeit usually justified ones, to the movement of goods. The Code never enjoyed a large number of signatories and that fact, coupled with its weak dispute settlement provision, resulted in no cases ever being brought under its provisions.

When the Uruguay Round Negotiations began in 1987, a number of nations including the United States decided to negotiate, for the first time, binding agreements on Technical Barriers to Trade (TBT) and on Sanitary and Phytosanitary Measures (SPS).

Both agreements are concerned with the avoidance of trade protectionist measures in the creation, use and enforcement of product standards, technical regulations and conformity assessment procedures. However, the agreements go about reaching this goal in quite different ways. The TBT relies on nondiscrimination - asking the question as to whether the measure results in discrimination between an imported product and the same or like products manufactured domestically or imported from another Party. Clearly such a test would not work well for SPS measures since such measures are designed to protect human, animal, and plant life and health at the level of protection that a Party deems appropriate. To achieve this kind of safety protection, nations must often discriminate against imported goods and among imported goods of different foreign countries since the origin may be a key factor in different risks of plant or animal pest or disease. Thus, the SPS Agreement differed from its inception from the TBT since it focuses on whether a measure has a basis in science and is based on a risk assessment, where appropriate. The precautionary principle is preserved since countries, in the absence of data relevant to risk, are free to adopt provisional measurers until such time as sufficient data are available to make a risk assessment (SPS Article 5.7).

Both TBT and SPS Agreements require the use of international standards with certain exceptions. For example, TBT Article 2.4 exempts instances where international standards would be ineffective for fulfilling legitimate objectives. SPS Article 3.3 exempts when the international standard does not meet the level of protection of human, animal or plant life or health a member country deemed appropriate.

For SPS measures countries may discriminate among products in trade so long as the discrimination is not arbitrary or unjustifiable. Nor shall a measure constitute a disguised restriction on international trade (SPS Article 2.3).

What is a SPS Measure?

Intent plays a critical role in the determination of what constitutes a SPS measure. In order to be a SPS measure it must be intended to:

1. protect animal, plant life, or health from risks arising from the entry, establishment, or spread of a pest or disease;

2. protect human or animal life or health from risks arising from an additive, contaminant, toxin, or disease-causing organism in a food, beverage or feedstuff;

3. protect human life or health from risks arising from a disease or pest carried by an animal, plant, or a product thereof;

4. protect or limit other damage from the introduction, establishment, or spread of a pest.

All of the above risks must be to human, animal or plant life or health "in the territory of the member" (emphasis added) in order to fit this definition. Thus, a threat to life or health occurring wholly in the global commons would not constitute a justification for an SPS measure under the WTO SPS Agreement. Countries are, of course, free to seek international solutions to such international problems. Allowable levels of pesticide residues on imported foods or beverages would fall under the section 2. above, allowing action to protect against risk from contaminants. The Food, Drug, and Cosmetic Act (FDCA) defines contaminant to include pesticide residues and gives EPA the authority to establish allowable levels and FDA the authority to enforce those levels in the United States.

SPS measures include end product criteria, process and production methods, testing, inspection, certification or approval procedures, methods of risk assessment, packaging and labeling requirements directly related to food safety, and quarantine treatments.

The North American Free Trade Agreement (NAFTA)

NAFTA (*3*) was agreed prior to the final negotiations on the WTO agreements. However, an inter0im draft text of the WTO was used as a basis for the NAFTA negotiations. NAFTA involved only three countries and was able to deal more explicitly with areas of concern in the health, safety and environment Thus the text, which all three countries have stated is fully consistent with the WTO agreements, contains similar but not identical obligations to avoid certain types of discrimination and to utilize science. It also contains interesting elaborations of such areas as the requirement to use international standards. NAFTA makes clear that there will be no "downward harmonization" of SPS measures. Thus, while harmonization of standards is encouraged it is clear that this cannot be to a less stringent standard. While governments are to use international standards as a basis (but not the only basis) for their SPS measures, Article 713(1) states that this is to be done "without reducing the level of protection of human, animal or plant life or health". Article 713(3) provides that nothing in this requirement "shall be construed to prevent a Party from adopting, maintaining or applying, in accordance with the other provisions of this Section, a Sanitary or Phytosanitary Measure that is more restrictive than the relevant international standard, guideline or recommendation."

Proposed Fast Track Authority

The U.S. Administration unsuccessfully sought renewal of fast-track in order to conclude additional trade agreements while " ensuring health, safety and environmental protection." Although fast track was not granted by the House of Representatives in 1997, it is useful to know what was put forward regarding environmental concerns. The

Administration laid out its environmental position related to fast track in a "Statement of Executive Initiatives Accompanying Fast-Tract Legislation of November 3, 1997." The Administration stated, "Even as we use this authority to negotiate strong new trade agreements, we need to ensure that our trade agenda complements and reinforces other important policy objectives, as it always has done. These goals include helping promote greater attention by other countries to the protection of the environment and worker rights." Specifically, the Administration stated its intention to strive for greater transparency and openness in the dispute settlement. It proposed that the WTO establish a group of prominent environmental experts to examine issues and formulate opinions for WTO to consider in such areas as the relationship between WTO rules and trade measures in international environmental agreements and ecolabeling. In addition, the Administration proposed to expand efforts to get international financial institutions to incorporate environmental issues into their operations as well as a greater flow of resources to environmental projects. These issues and possibly others will need to be readdressed whenever the Administration resumes its efforts to obtain fast track authority. The possibility of getting congressional approval in 1998 is not great.

The President's New Transatlantic Initiative

In 1996 President Clinton agreed with the President of the European Union to create a new transatlantic initiative which would be industry driven and aim at the removal of barriers to trade between the United States and The European Union through a series of "Mutual Recognition Agreements". In 1997 this cooperative effort resulted in the text of an "Agreement on Mutual Recognition between the United States and the European Community". The package, which was initialed by USTR and EC representatives in 1997 but not yet formally adopted by the governments, consists of an umbrella agreement and 5 sectoral agreements in telecommunications, electrical safety, medical devices, recreational craft, and pharmaceutical good manufacturing practices (see www.ustr.gov). Critics of these agreements charge that they were not developed in an open and transparent manner, and that bunching several sectors together puts unnecessary pressure on health, safety and environmental protection. They point to the fact that the EU refused to sign the MRA on telecommunications until agreement was reached on medical devices and pharmaceutical good manufacturing practices. The allegation is that by employing tactics which are quite common in trade negotiations (e.g. linking tariff reductions in one sector to reductions in a totally different sector) the EU is placing unprecedented pressure on health and safety issues. New areas currently under discussion by relevant sectors of industry include: pharmaceuticals, industrial chemicals and biotechnology. One of the areas in which industry has signaled it wishes to see mutual recognition agreement is good laboratory practices.

Long time advocates of the OECD GLPs, while happy to see that the OECD principles and compliance mechanisms are to be the core of the agreement, question whether a separate bilateral between the EU and the U.S. is necessary in view of the recent revisions to the OECD procedures which provide for joint inspections.

Industry points to the fact that for whatever reason there has not been proper application of the OECD Council Decision on Mutual Acceptance of Data and they believe the MRA will serve to remove remaining impediments to tests done on one side of the Atlantic in accordance with OECD GLPs being accepted for assessment by the regulatory agencies on the other side.

Meeting the Level of Health, Safety and Environmental Protection an Importing Country Deems Appropriate

Both the SPS and the TBT Agreements require Parties to use international standards under certain conditions and grant a presumption of consistency to the WTO agreements if international standards are used. However, they both contain exceptions to the use of such standards. The TBT specifies that if a Party can show that international standards are an ineffective or inappropriate means for the fulfillment of the legitimate objectives pursued then international standards need not be used as the basis for technical regulations. The agreement explicitly recognizes protection of health or the environment as legitimate objectives.

The SPS allows SPS measures which result in a higher level of protection than those arising from an international standard if there is a scientific justification, or as a consequence of the level of sanitary or phytosanitary protection a Member determines to be appropriate.

What Constitutes Appropriate Procedures for Good Laboratory Practices Related to Health and Environmental Testing?

Specifically, do the WTO Agreements and NAFTA require the use of ISO/IEC Guide 25 by laboratories generating data for regulatory purposes? After all, ISO is explicitly recognized in the trade agreements as an international standard setting body. The answer is that only if countries determine that ISO/IEC Guide 25 will meet the criteria of the SPS Agreement would adherence be required. Under that agreement, Parties may use a SPS measure which results in a higher level of sanitary or phytosanitary protection than would be achieved by measures based on the relevant international standards or guidelines if there is a scientific justification or as a result of the level of sanitary or phytosanitary protection a Member determines to be appropriate. It is important to note that a government may establish its levels of protection by any means available under its law, including by referendum. The choice of the appropriate level of protection is a societal value judgement. The agreement imposes no requirement to establish a scientific basis for the chosen level of protection because the choice is not a scientific judgement.

Both FDA and EPA have for many years been actively involved in the development of GLPs under the auspices of the Organization for Economic Cooperation and Development (OECD). The OECD Member Countries have decided at a policy level that "the driving force for GLP compliance monitoring is the requirement to assure regulatory authorities that data they receive in safety studies can be relied upon when making assessments of hazards or risks to man, animals and/or the environment."

This was made clear in "*The Position of the OECD Panel on Good Laboratory Practice as endorsed by the 22 Joint Meeting of the Special Program on the Control of Chemicals, November 16, 1994.*" The accuracy of such data is critical to achieving the level of protection of the environment, health and safety that a country deems appropriate.

A laboratory accredited according to ISO/IEC Guide 25 (Guide 25) to carry out specified determinations on a continuing basis in areas of physical, chemical, and analytical procedures may satisfy many of EPA's and OECD's GLP requirements. However, certain fundamental requirements of the OECD GLP Principles are not covered by laboratories accredited according to Guide 25. Deficiencies in Guide 25, according to the OECD Position, include a lack of study plans and any requirement to designate a Study Director. Other OECD requirements such as recording and reporting of data, management of retained data to allow complete reconstruction of a study, and a program of independent quality assurance are more stringent under OECD GLPs as well as the GLPs of EPA and FDA. There are strong scientific justifications to support these fundamental requirements. For this reason, data generated solely under ISO/IEC Guide 25 are unlikely to be accepted by regulatory authorities of the OECD Member Countries for purposes of assessment of chemicals related to protection of health and the environment.

This does not mean that the movement toward the use of international standards should be ignored. Although OECD is now in the process of expanding its longtime membership of 25 countries to include some of the important emerging economies and countries in transition, OECD by its nature cannot itself qualify as an international standard setting body since the parent body is not and has no plans to be open to participation by all parties to the WTO agreements. It is interesting to note that a number of non-OECD countries including Brazil, India and China (which is not a member of the WTO or a Party to any of its agreements) have indicated their interest in being associated with the OECD Decision on Mutual Acceptance of Data (MAD). That Decision binds OECD countries to accept, for purposes of assessment, data generated according to OECD test guidelines and in compliance with the OECD GLPs. A new Decision to expand the MAD beyond OECD to meet this growing interest of non-OECD countries to participate was adopted by the OECD Council in November 1997.

Voluntary Consensus Standards

In the United States, in addition to being required by the trade agreements to use international standards under certain conditions as outlined above, federal regulatory agencies must also comply with a recent law requiring the use of voluntary consensus standards (*4*) in certain circumstances. The enactment, in 1996, of the "National Technology Transfer and Advancement Act of 1995" (NTAA), Public Law 104-113 (NTAA), specifically Section 12(d), directs federal agencies, in their regulatory and procurement activities, to use voluntary consensus standards in place of standards created by and unique to the federal government except where the use of voluntary consensus standards would be inconsistent with U.S. law or otherwise impractical. The new statutory requirements in many ways codify an earlier Executive Branch directive from the Office of Management and Budget (OMB Circular No. A-119, issued October 20, 1993). OMB published a final revision to Circular A-119 on February 19, 1998 to make it consistent with the new law. The primary purpose of the new revision is to interpret the provisions of Section 12 (d) so that federal agencies can properly implement the statutory requirements. These include the requirement that all federal agencies use domestic or international voluntary consensus standards instead of government-unique standards in their procurement and regulatory activities. If the agency uses a government-unique standard in lieu of a voluntary consensus standard it must provide an explanation to OMB, which in turn, reports to Congress.

Although the law does not mention international standards, the Circular clarifies that requirement extends to international as well as domestic voluntary consensus standards. It is clear that some "international standards" (as that term is used in WTO agreements) could meet the definition of "voluntary consensus standards" (under NTTAA). Indeed voluntary consensus standards are standards developed or adopted by domestic or international voluntary consensus standards bodies. The key is whether the standard is truly consensus based. In making a determination whether this is so, it will be important to examine the process by which the standard was created to ascertain if that process was open and transparent, contained a proper balance of interests, followed the concept of due process including allowing for some form of appeal procedure, and had the general agreement of concerned parties. Because of these criteria, there are and will be standards developed by the private sector that will not qualify as voluntary consensus ones. These include company or industry standards developed without a consensus process. For example, questions have been raised as to whether the "canvas method" of standards development meets the tests of openness and due process. The determination of whether a particular domestic or international standard qualifies as a voluntary consensus standard is to be made by the agency that has the responsibility for implementing the laws which it is charged to uphold.

The word "international standard" does not appear in NTTAA. However the "Policy for Federal Use of Standards"contained in item 6.h of Circular A-119 clarifies that the policy does not establish a preference between domestic and international voluntary consensus standards. It also notes that "in the interest of promoting trade and implementing the provisions of international treaty agreements, [federal] agencies should consider international standards in procurement and regulatory applications." This oblique reference is useful since both the WTO and NAFTA trade agreements require the use of international standards (with certain important exceptions).

Can the OECD Test Guidelines and Good Laboratory Practices qualify as voluntary consensus standards under the recent amendments to the NTTAA? They were developed in an open and transparent manner with active participation by industry, governments and the environmental communities. However OECD does not consider itself to be a voluntary consensus standards body. It will be important to record use of these guidelines in the annual report to OMB and Congress since FDA, EPA and the private sector make extensive use of OECD guidelines.

To identify whether voluntary consensus standards exist that are relevant to a particular regulation, agencies will utilize a number of data bases including those maintained by ANSI, standards bodies, and standards publishing companies as well as a library maintained by the National Institute of Standards and Technology (NIST),. NIST is required under the new law to coordinate Federal, State, and local standards activities as well as conformity assessment activities with private sector activities in the same field. The Secretary of Commerce is developing guidelines for the agencies.

Agencies are required to report through NIST to OMB, on an annual basis, the use of voluntary consensus standards and explain why in cases where government unique standard is used instead of relevant, existing, voluntary consensus standard. Agencies will also ensure that there is proper notice in Notices of Proposed Rule Making or Interim Final Rules regarding the use or non-use of voluntary consensus standards. These steps will help to ensure the openness of the process and allow all interested parties to comment on the availability and appropriateness of federal use of a particular voluntary consensus standard.

Please note that the new law does not limit the authority and responsibility of agencies to make regulatory decisions about the level of acceptable risk or the level of health, safety, or environmental protection that the agency determines to be appropriate. While the new law establishes no preference between domestic and international voluntary consensus standards in order to implement the provisions of the international trade agreements, EPA and other agencies will consider the possible use of relevant international standards in their procurement and regulatory actions. This approach is in the interest of promoting trade and of implementing the provisions of the international trade agreements.

Conclusion

The requirements on federal agencies stemming from the Trade Agreements and from the requirements arising from the National Technology Transfer and Advancement Act are not identical, but they are not necessarily contradictory either. To administer them wisely in a consistent manner and to do so while continuing to fully protect the environment, safety, and health will require both an increase in knowledge and intelligent application of that knowledge to all future regulations. The best hope of this occurring is to ensure and rely on an open and transparent process which involves the private sector, academia, and public interest groups. To the extent that standard- setting bodies can work to ensure that all standards relating to trade are not only voluntary consensus ones but also fit the definition of international standards, the lives and choices of the bureaucrats will be made simpler and the results more beneficial to all.

Literature Cited

1. The Final Act embodying the Results of the Urugruy Round of Multilateral Trade Negotiations, April 15, 1994. Reprinted by the World Trade Organization (WTO), ISBN 92-870-1121-1 1995, and on the WTO Home Page, www.wto.org.
2. President's Message to Congress transmitting the Uruguay Round Trade Agreements, Texts of Agreements Implementing Bill, Statement of Administrative Action and Required supporting Statements, H.R. Document 103-316, Vol. 1. Sept. 27, 1994.
3. President's Message transmitting the North American Free Trade Agreement, Texts of Agreement, Implementing bill, Statement of Administrative Action and Required Supporting Statements, H.R. Document 103-159 Vol. 1. November 4, 1993.
4. Federal participation in the Development and Use of Voluntary Consensus Standards and in Conformity Assessment Activities, OMB Circular A-119, 63 Fed. Reg. 8546 (February 19, 1998).

Chapter 3

International Harmonization via ISO

Peter Spath

Eastman Kodak, 1669 Lake Avenue, Rochester, NY 14652–4603

Accreditation, certification, and registration all have different definitions dependent upon location and process. Although these terms have a different connotation and purpose they do point toward a common goal, consistency. The ISO standards can provide that consistency when properly implemented. ISO provides standards for certification of the laboratory, auditors, and quality systems. This chapter will address the relevant standards and the intended uses.

The ISO 9000 series of standards provides guidance for quality management and models for quality assurance by describing what elements quality systems should encompass. As in the GLPs, ISO also emphasizes the quality audit as an important tool for achieving key management objectives.

ISO/IEC Guide 25 (*1*) provides a mechanism to implement a quality system in the laboratory that satisfies all the relevant requirements of the ISO 9000 standards. The next edition of ISO/IEC Guide 25, currently in draft form, more closely resembles the GLP Regulations, thereby providing a good backbone to support a GLP compliant system.

The ISO standard that is currently receiving much attention is the new ISO 14000 series of Environmental Management Standards. Of particular interest is ISO 14001 (*2*), the one standard in the series that carries a certification step for Environmental Management Systems. Like the ISO 9000 standards, this will also be voluntary, but it is anticipated that pressure from the market and, in particular, environmental groups will force industry down the path to certification/registration.

The standards of the International Organization for Standardization (ISO) provide guidance for an organization to develop management systems that can be implemented in areas with existing regulatory programs. They provide a versatile structure allowing an organization to integrate regulatory, customer and its own needs into one management system. The competition in today's global market has forced organizations into a mode of continuous improvement to meet the customer's ever increasing expectations. In order to compete and still maintain good economic performance, organizations have had to employ more effective and efficient quality management systems. The ISO 9000 series of standards provides guidance for quality management and models for quality assurance by describing the elements an effective quality system should encompass.

It is the development of an effective system that an organization should focus on initially. To choose the most appropriate system for the organization, first turn to ISO 9000-1, *Quality Management and Quality Assurance Standards - Guidelines for Selection and Use (3)*. This standard is the road map for the entire ISO 9000 family. It explains the roles of each of the of ISO 9000 standards, discusses concepts that are basic to the entire series, and will help identify the key objectives for quality of product and processes.

Once an organization's objectives are clearly defined, ISO 9004-1, *Quality Management and Quality System Elements - Guidelines* (*4*), should be consulted next. This often overlooked document describes the elements necessary for an effective quality management system. Any organization will receive significant benefit from implementing and using a quality system based solely on this standard. It is felt that a quality system based on this standard should be used on a daily basis to help satisfy customer needs and expectations, while ensuring that an organization's own needs are served. This standard outlines the various aspects of management's responsibility for the quality system and how those objectives should be documented. The elements of a quality system are explained in detail, rather than just stating the requirement as is done in the contractual standards. There are also guidelines for elements not included in the rest of the ISO 9000 series that should be in any good quality system, such as, guidelines for providing quality in the marketing and purchasing functions; assistance in defining the specification, design and review process; and even ideas on how to gather, present, and analyze the elements of financial data.

When it comes to defining contractual requirements between a supplier and a customer, other standards in the 9000 series become useful. These standards provide a subset of the requirements in ISO 9004-1 to which an organization should demonstrate compliance when entering into a contract with a supplier or customer. The contractual standards are becoming well known as:

ISO 9001:1994 - *Model for Quality Assurance in Design, Development, Production, Installation and Servicing* (*5*)
ISO 9002:1994 - *Model for Quality Assurance in Production, Installation, and Servicing* (*6*)
ISO 9003:1994 - *Model for Quality Assurance in Final Inspection and Test* (*7*)

Whichever standard is chosen, the key to effective implementation of an ISO compliant quality system is to have full management commitment. This should be two fold. The first commitment is to provide all the necessary resources for effective development and implementation of the quality system. After implementation, management should be actively involved to assure that the system continues to survive and improve. Without continuous improvement, regardless how small, most systems will stagnate and eventually just be another manual on a shelf.

Compliance to the standard can be demonstrated through either verification by customers, or formal registration of the quality system with independent third party registrars. Audits by customers will usually focus on specific products or contractual requirements, whereas registration provides evidence that a company's quality system satisfies the requirements contained in one of the ISO 9000 quality assurance standards.

ISO 9000 was founded on the premise that all work is accomplished by a process. All processes have inputs and outputs, otherwise known as products. In accordance with the ISO definition, a product can be tangible or intangible. Products may include hardware, services, knowledge, or concepts. These standards are written in generic terms of quality system objectives which need to be satisfied. They do not prescribe how to achieve the objectives. Those objectives, and the manner in which they can be achieved, will be different for each organization that implements a quality system. An organization that manufactures stuffed animals will be vastly simpler than an organization that provides health or environmental data to a Federal Agency.

The US Food and Drug Administration (FDA) felt that a significant portion of medical device failures were due to poor design controls. In their efforts to include design controls in the current Good Manufacturing Practices (cGMP), ISO 9001 was used as a model. On June 1, 1997, the FDA cGMP for medical device manufacture became the Quality System Regulation. This change allows organizations, where only a portion of their production concerns medical devices, to implement one quality system based on the Quality System Regulation. Whereas, they previously had an ISO 9001 system for the manufacture of the majority of their products and a cGMP system for the medical device manufacture. Having two different sets of requirements for record keeping obviously lead to frequent problems.

The ISO standards are typically on a 5 year revision cycle, although the original versions of the ISO 9000 series, published in 1987, were not revised until 1994. The ISO technical committee, TC 176, has formally established the specifications for the next revision of ISO 9000, which is supposed to be published in 1999. ISO 9001, 9002, and 9003 will become a single standard to be numbered ISO 9001. ISO 8402, *Quality Management and Quality Assurance -Vocabulary,* (*8*) will become ISO 9000 entitled concepts and terminology. ISO 9004 will continue to be the guidance document for the ISO 9001 standard, and will address quality management systems in general.

ISO/IEC Guide 25, *General requirements for the competence of calibration and testing laboratories,* was developed in collaboration with the International Electrotechnical Commission to provide a mechanism to promote confidence in calibration and test laboratories worldwide. This confidence is gained by proving that a laboratory can operate in accordance with the requirements of this guide as documented by the lab. By providing a system that promotes consistency between laboratories, states, federations, and countries, harmony can be achieved.

Members of ISO or IEC participate in the development of international standards through technical committees that deal with specific technical activities. When Guide 25 was in development, particular attention was paid to other requirements for laboratory competence such as those laid down in the OECD Principles of Good Laboratory Practices and the ISO 9000 series of quality assurance standards. Laboratories complying with the requirements of Guide 25 satisfy the relevant requirements described in the ISO 9000 standards for a service organization.

The implementation of quality systems in the laboratory has greatly increased since Guide 25 was published in the early eighties. Many industries have adopted it as the basis for establishing consistent quality management in a laboratory. Recognizing the competence of a laboratory can be done by awarding accreditation through either an internal or third party accreditation processes. The use of third parties to award accreditation to Guide 25 is gaining popularity in the United States. Registrars not only provide technically oriented facility inspections, but also proficiency sample programs where required.

The EPA has recently integrated ISO standards into a new regulation with the introduction of the National Environmental Laboratory Accreditation Program (NELAP). Prior to this program each state had its own requirements for the accreditation of laboratories involved in environmental testing. This required multi-site organizations to maintain multiple quality systems to satisfy the different requirements of each jurisdiction in which they do business. Not only were there fees for each state program in which they were enrolled, but the audits conducted by each state on at least an annual basis became extremely costly. Some large organizations have full time staff dedicated to escorting auditors and responding to audit reports. ISO/IEC Guide 25 was integrated into a national standard for environmental laboratories so that each facility has only one quality system to maintain regardless of how many states they conduct business in. Reciprocity between states also eliminates the "audit of the week" situation by placing the responsibility for conducting the on-site audit with the state in which the facility is located. Obviously, by eliminating multiple audits in the laboratory, harmony is promoted within the organization too.

The ISO standard that is currently receiving much attention is the new ISO 14000 series of Environmental Management Standards. Of particular interest is ISO 14001, *Environmental Management Systems - Specification with Guidance for Use* which was published in September 1996. This is the only standard in the series that carries a certification step for Environmental Management Systems. In most developed countries, it was the standards-setting organizations that pushed the implementation of this standard, whereas industry was the primary participation group in the United States. Although many took a wait-and-see attitude in the beginning, registration has been gaining momentum recently. Like the ISO 9000 standards, registration is voluntary, but it is anticipated that pressure from the market and environmental groups can force industry down the path to registration. Whether organizations choose to go through the process of third party registration or not, ISO 14001 is an effective means to streamline operations, prevent pollution and reduce costs.

This international series of environmental management system standards is intended to guide organizations in managing the immediate and long term impact that its product, services and operations have on the environment. Effective implementation of ISO 14001 should help an organization anticipate and meet growing environmental impact expectations by incorporating those issues into the business planning process. Hopefully, it will also ensure that an organization can maintain ongoing compliance with national and international requirements while still being competitive in the global marketplace.

As important as what is covered in this standard is what is not included. The goal of ISO 14000 is registration to an auditable standard by focusing on consistency within the system. It does not specify test methods or limit levels for pollutants, nor are quantified performance levels prescribed.

The most inconsistent variable in the ISO accreditation process can be the auditors. To provide consistency, certified auditors should be used for conducting quality systems audits. Certification of auditors can be accomplished through a few different avenues. The International Auditor Training and Certification Association (IATCA) program is formed of national organizations that offer accreditation of auditor training courses and certification of quality systems auditors. Another program is the US Registrar Accreditation Board (RAB) which developed and administers programs to both certify auditors and accredit courses for training auditors. This affiliate of the American Society for Quality (ASQ) provides internationally recognized certification designed to assure that auditors possess the qualifications required to audit quality systems using a variety of standards, including the GLPs. The FDA has recently contracted with the RAB to develop a certification program for cGMP Auditors using the ISO standards for the program criteria.

To qualify for certification, a candidate must first satisfy basic requirements concerning education, training, work experience, and personal attributes. Once these basic requirements have been met, a candidate may apply for an entry level grade, such as quality systems provisional auditor. This grade level is for those with little or no quality systems auditing experience, but does allow a person to participate on audits and gain the experience needed for advancement of grade. While a program such as this allows for direct application to any grade, it also provides the opportunity to advance in grade as experience is gained.

A candidate for Quality Systems Auditor must demonstrate the ability to participate effectively on audit teams during actual audits. The required number of audits for certification as an auditor varies with the educational level and work place experience of the applicant.

Candidates for Quality Systems Lead Auditor must, in addition to demonstrating the ability to participate on an audit team, demonstrate the ability to manage and lead audit teams during actual audits. In order to apply for this grade, all audit experience must be witnessed by a certified Quality Systems Lead Auditor.

All certified auditors must maintain their auditing skills through ongoing auditing experience and continued education. The audit and educational experiences must be documented on logs which are then supplied as part of the Annual Application for Continuation of Certification.

To harmonize the manner in which audits are conducted, the guidelines provided in ISO 10011 (*9-11*) have been accepted internationally as the standard for auditing quality systems since 1991. The three parts of this standard include auditing, qualification criteria for quality systems auditors, and management of audit programs. As in many regulatory compliance programs, ISO emphasizes internal quality system audits as an important tool for achieving key management objectives. A company can utilize an internal auditing organization to provide independent assessments of their quality system. An audit of this type would provide assurance that the quality system implementation has been effective and will also satisfy the requirements for an internal audit if third party accreditation is sought. For many organizations an internal compliance program may be sufficient. However, for a large organization competing in the global marketplace, registration is often considered a necessity to achieve the recognition.

In a GLP regulated facility, study specific audits are required and are the responsibility of the Quality Assurance Unit (QAU). Such specific requirements would be integrated into the quality system using this series of standards to organize and manage the required aspects of the QAU. In this situation, the QAU would conduct the vertical audits assuring that the specific study complies with all aspects of the GLPs and relevant sections of ISO/IEC Guide 25. The registrar would then conduct audits of the organization including the QAU to assure that the internal program is effective and capable of performing its function in accordance with ISO/IEC Guide 25 and GLPs as required by clause 12.1 of the Guide.

The effectiveness of corrective actions taken by the organization in response to internal audits must be assessed by the registrar. Findings of an internal audit in a regulated environment should be kept confidential to protect the organization. One way to maintain this confidentiality is to conduct these audits under Attorney-Client Privilege.

However the effectiveness of the quality system is assessed, its success depends on commitment at all levels, especially from the highest levels of management. A good quality system enables an organization to establish and assess the effectiveness of procedures which set policies and objectives, achieve conformance to them, and demonstrate such conformance to others.

Although the terms accreditation, certification, and registration have different connotation and purpose, they do point toward a common goal, consistency. It is through consistency that harmonization can be achieved. Seeing these terms in context throughout this paper may help differentiate between them. Since there is so much confusion surrounding these terms, the following Table will help explain them:

Table 1. Assurance of Conformity

	LABORATORY	QUALITY SYSTEM	PRODUCT
U.S.	ACCREDITATION	REGISTRATION	CERTIFICATION
ELSEWHERE	*ACCREDITATION*	CERTIFICATION	CERTIFICATION

We can see that different terms are used for different purposes in different places. Even if the word is the same, such as accreditation, the connotation is different. Laboratory accreditation in the United States is actually closer to what is considered certification in other countries. We have varying degrees of accreditation in the United States, some of which are much more rigorous than others. Under some accreditation schemes, performance evaluation samples are required to assess the technical capability of a facility while others rely on thorough, technically-oriented facility inspections. The greatest difference is that accreditation in the United States is provided, most often, by private organizations whereas in other locations it is primarily government-related bodies that have the authority. One exception is NELAP, which is administered under the US EPA and it is the state governmental bodies that will be registered as the accrediting authorities.

Quality systems in the United States become registered when they have been assessed by a third party as conforming to an accepted set of specifications. Elsewhere quality systems are certified, once again, by government-related organizations. But, the term registration is gaining acceptance worldwide. This should help to alleviate the confusion that has existed, since in most locations certification implies compliance to both product-specific technical requirements and quality system requirements.

Product certification is nearly the same across international boundaries with the main exception being that in the United States certification comes from private organizations, whereas elsewhere it is primarily a function of government-related organizations. It is this difference that has created the great debate between the United States and the rest of the world. We understand the current position of the European Accreditation of Certification (EAC) is that in order to be trustworthy, accreditation must not be conducted in a competitive environment. Competition is not permitted in Europe, but the European accreditation bodies actively compete to accredit United States based laboratories.

Other than the US EPA, in areas such as NELAP, there is only one system in the US for accrediting quality system registrars, the RAB, and there is only one system for accrediting product certifiers, ANSI. Both of these are private sector organizations. We do not have any government appointed bodies to provide accreditation, registration, and certification. In the true American style we have competition amongst the accreditation organizations. There are more than 55 registrars currently operating in the United States. Companies are free to pay a registrar of their choice for the registration of their quality system. Depending on the size of the company, this cost will be anywhere from 5,000 to 35,000 US dollars. But this is only about 10% of the total cost of quality system implementation. On average companies spend about 75% on internal efforts and another 15% on external resources such as training and consultants over about 18 months.

Hopefully, I have provided some ideas on how to achieve harmonization amongst the various management systems within your organizations. Whether those systems exist to meet customer or regulatory requirements, a quality system based on the ISO standards can help harmonize your internal programs. And once we have all achieved internal harmony it will naturally carry over to the big picture - global harmonization.

Literature Cited

1. ISO/IEC Guide 25:1990, *General requirements for the competence of calibration and testing laboratories.*
2. ISO 14001:1996, *Environmental Management Systems - Specification with Guidance for Use.*
3. ANSI/ASQC Q9000-1-1994, *Quality Management and Quality Assurance Standards - Guidelines for Selection and Use.*
4. ANSI/ASQC Q9004-1-1994, *Quality Management and Quality System Elements - Guidelines.*
5. ANSI/ASQC Q9001-1994, *Quality Systems - Model for Quality Assurance in Design, Development, Production, Installation, and Servicing.*
6. ANSI/ASQC Q9002-1994, *Quality Systems - Model for Quality Assurance in Production, Installation, and Servicing.*
7. ANSI/ASQC Q9003-1994, *Quality Systems - Model for Quality Assurance in Final Inspection and Test.*
8. ISO 8402:1994, *Quality Management and Quality Assurance -Vocabulary.*
9. ANSI/ASQC Q10011-1-1994, Guidelines for Auditing Quality Systems - Auditing.
10. ANSI/ASQC Q10011-2-1994, Guidelines for Auditing Quality Systems - Qualification Criteria for Quality Systems Auditors.
11. ANSI/ASQC Q10011-3-1994, Guidelines for Auditing Quality Systems - Management of Audit Programs.

Chapter 4

Laboratory Competence: ISO Guide 25 or GLP?

John Gilmour and Helen Liddy

National Association of Testing Authorities, Australia, 7 Leeds Street, Rhodes, NSW 2138, Australia

Both the OECD Principles of Good Laboratory Practice (GLP) and ISO/IEC Guide 25 (G25) were developed in the late 1970s in response to the need to eliminate technical barriers to trade, one of the most significant of which is lack of acceptance of test data. GLP is applicable only to testing of chemicals while G25 seeks to address generic laboratory practices for use in any situation but particularly for industrially manufactured products. Both documents have much in common but with differences in emphasis, application terminology and content. The paper will examine these differences and also consider current trends in some countries to use the techniques of accreditation (against GLP) for monitoring purposes. It will also look at the relevance of the international quality systems standards ISO 9001 & 9002 in the context of laboratory performance.

There is widespread international interest in the quality and integrity of laboratory test data. This interest stems from a variety of pressures, but most notably it is related to questions of health, safety, the environment and trade.

Testing is expensive and it is highly desirable to develop arrangements whereby testing carried out competently in one country can be accepted in other countries. To achieve this, users of test data, whether they be regulatory authorities or commercial customers, require confidence in the competence of laboratories generating data. Hence, they often also need confirmation of that confidence in specific sets of data.

There has, therefore, been considerable international effort during the past twenty years to address these needs in both general terms and for specific areas. From this work there are two prominent outcomes that have much in common but they also have some significant differences in emphasis. Each is used within the OECD countries that have both strong trading economies and a particular concern with assuring the quality and integrity of laboratory data utilized to assess the risks to humans and the environment from chemicals or chemical products. The two systems in question involve:

(a) the application of OECD Principles of Good Laboratory Practice (GLP) (*1*) by laboratories testing chemicals, enforced by monitoring authorities (usually, but not always, government) that assess compliance of laboratories with GLP; and

(b) laboratory accreditation systems applied more generally in a trading context that have their basis in ISO/IEC Guide 25 (*2*). Such systems are also often, but not exclusively, administered by government or quasi-government bodies.

There is overlap between the two approaches and some laboratories are required to satisfy both, which causes a degree of inconvenience, cost and confusion. Lack of understanding within the user community has also led to confusion and misunderstanding as to what can be expected by the application of either system.

This paper examines the differences in emphasis, application, terminology and content between the two systems, and also considers current trends in some countries to use the technique of accreditation (against GLP requirements) for monitoring compliance.

Origins and Purposes of the Systems

The OECD Principles of Good Laboratory Practice were developed for use within the OECD to provide an international understanding among member countries and to give a framework within which test data, in the specific area of hazardous chemicals, could be accepted internationally, thereby avoiding a serious non-tariff trade barrier. The purpose was, thus, to avoid any suggestion that the quality of test data could be an issue in the international exchange of that data when used for regulatory purposes.

Laboratory accreditation was developed to give any user of laboratory test data confidence in the competence of a particular laboratory and thus to enhance confidence in specific sets of data. It was originally developed solely for internal application within a few countries but, through networks of mutual recognition agreements, is now widely used for trade.

While the widest use of laboratory accreditation remains in domestic trade and commerce, there is considerable international interest in using it to avoid the use of testing generally as a non-tariff barrier.

Concepts

Laboratory Accreditation. The concept of laboratory accreditation was developed immediately after World War II to address a need for identification of competent laboratories working in more traditional and routine environments with a strong emphasis on calibration of instruments.

From the outset, laboratory accreditation was concerned with assessing the technical competence of laboratories working in any area of testing or measurement. It depended on the identification and codification of the elements regarded as necessary to produce reliable test results. Laboratories judged to satisfy the code were regarded as being technically competent and having a suitable quality system in place, and hence producing reliable data.

The supporting documentation which was developed paid close attention to the technical qualifications and skill of individuals working in a laboratory combined with the resources available to conduct specific tests. Attention was also given to matters which were defined as being "good laboratory practice" but were described in fairly general terms.

As the practice of laboratory accreditation has been truly internationalized, its documentation has retained much of the earlier technical specificity, but it has been greatly expanded to take into account the management emphasis to be found in the GLP documents and other systems type standards.

The aim in laboratory accreditation has been to provide confidence, in advance, to a prospective client or sponsor that a laboratory or other testing facility is capable of producing reliable study results or test reports on a continuing basis, usually in terms of specific methods of test. Laboratories seeking accreditation usually do so to provide confidence in the competence of the laboratory to its proprietor and its customers. Often larger purchasers, such as government procurement agencies, require accreditation as a prerequisite for contract work. Laboratory accreditation is also useful as a marketing tool for laboratories.

The technical scope of laboratory accreditation is much broader than that of GLP. It was developed to cover all testing and measurement situations. Laboratory accreditation organizations may specialize in specific fields but usually offer comprehensive programs covering sciences and technologies as diverse as acoustics, biology, chemistry, engineering, physics, and human and veterinary pathology.

The international generic standard used in laboratory accreditation is ISO/IEC Guide 25 which covers the general requirements for laboratories. All accreditation bodies must then develop specific criteria for the various areas of testing in which laboratories are likely to seek accreditation.

From the point of view of a laboratory accreditation body, the OECD Principles of GLP can be seen as one of this set of specific criteria which might broadly come within the fields of biology, chemistry, and human and veterinary pathology. GLP does, however, contain some criteria not specifically addressed in the international accreditation standards, such as animal management considerations.

Good Laboratory Practice. The Principles of Good Laboratory Practice and associated documents were produced specifically to meet the needs of a special group of regulatory authorities in OECD countries. The concern of these bodies is the validity of data produced by external laboratories but used by the authorities when making decisions about the safety and environmental impact of chemicals.

The important features are that the purpose of the system is to satisfy a particular set of clients with special statutory responsibilities, and decisions made by the authorities are not based on commercial consideration but rather on the level of potential hazard associated with the release of a product onto a particular market.

The authorities are, therefore, concerned with the scientific validity and integrity of specific sets of data.

On this basis, documentation was developed which described a management system which had a strong focus on medium to long term studies of chemicals for their effect on health and the environment. In recent times it has been used effectively by monitoring authorities for more general use and developed to include studies conducted in the field.

While Good Laboratory Practice (GLP) has been heavily oriented towards assessing the quality and integrity of completed studies on the health and environmental effects of new chemicals, previously untested (within the meaning of current practice) chemicals are now being exposed to full studies.

In GLP systems based upon the OECD Principles, each study submitted to a regulatory authority must be accompanied by a compliance statement by the laboratory, indicating whether or not the study was performed in accordance with GLP. It is, therefore, a self declaration by the laboratory as to its compliance with the Principles on a study by study basis. The monitoring authority audits such declarations. For laboratories carrying out regulatory studies for notification or registration of chemicals, the application of GLP is a *conditio sine qua non.*

For other work than that required to be performed under GLP, the regulatory authorities are not in any way concerned as to compliance by any laboratory with the Principles.

Because laboratories accredited under an accreditation program are usually expected to offer a more broadly based or continuing service to clients, laboratory accreditation bodies are concerned about each laboratory's continuing compliance with accreditation conditions, not just at particular times of special work.

The high level of involvement by the monitoring authorities has also led to the practice of individual study audits which are not a feature of traditional laboratory accreditation.

Scope of Application

Good Laboratory Practice. The scope of application of GLP is generally stated in terms of specific categories of products that are being regulated (i.e., subject to notification or registration requirements), and for which health and environmental safety testing is required as a condition of notification or registration. In fact, some regulatory systems specifically identify the types of regulatory studies that are subject to GLP and those that are not.

Although regulations vary from country to country, it generally can be stated that GLP applies to the notification or registration of the classes of chemicals or products below:

- industrial chemicals (usually all newly marketed chemicals and some specified chemicals already on the market);
- pharmaceuticals;
- veterinary drugs;
- pesticides;
- food and feed additives;
- cosmetics.

The types of safety studies to which GLP applies are generally:

- physical-chemical testing;
- toxicity studies;
- mutgenicity studies;
- environmental toxicity studies on aquatic and terrestrial organisms;
- studies on behavior in water, soil and air; bioaccumulation;
- residue studies;
- studies on effects on mesocosms and natural ecosystems;
- analytical and clinical chemistry testing.

Laboratories may specialize in one or a few limited areas or offer a comprehensive service.

Laboratory Accreditation. All national comprehensive laboratory accreditation bodies offer accreditation in terms of some combination of area of testing (chemical, biological, physical, etc.), product (foodstuff, pharmaceutical, concrete, steel, etc.) testing techniques (mass spectrometry, gas chromatography, IR, etc.) and usually analyte or analyte groups. A laboratory's competence is specified by some combination of these four elements.

For example, a laboratory may be accredited for:

- *chemical tests on waters*
- *analysis using classical, AAS, HPLC techniques by the methods of APHA and EPA*
- *the following determinations: acidity, dissolved solids, trace metals, etc.*

Requirements

For widespread international harmonization it is necessary that any system be transparent and be seen to be so, and that common standards and practices are applied.

For mutual recognition of test data these requirements must be met at two levels. First, the criteria against which laboratories are evaluated must be technically appropriate and, secondly, the policies, procedures, practices and techniques of evaluation used by accreditation or approval bodies must be fully documented and agreed by the parties to any agreement.

GLP requirements are set out in the OECD Principles of Good Laboratory Practice, and in national regulations derived from or equivalent to the GLP Principles.

Under laboratory accreditation programs, laboratories are expected to comply with the generic requirements of ISO/IEC Guide 25, and accreditation bodies themselves need to comply with standards such as ISO Guide 58 (*3*).

From the point of view of labor ıtories, the two systems contain many similar requirements. These concern the organization of a laboratory, its facilities, qualification of scientific and technical staff, documentation of procedures, record keeping and reporting.

Textual comparison of the requirements of the two systems has often led to confusion, due almost entirely to differences in terminology. The text of the OECD Principles of GLP is specifically concerned with the health and environmental safety testing of chemicals and uses terminology prevalent in that context, whereas the language of laboratory accreditation attempts to be more generally applicable.

A factor which is not always appreciated is that, while laboratory accreditation utilizes ISO/IEC Guide 25, this is only a generic document which requires development of specific technical documentation for each area of testing to be covered by the accreditation body, and this material is usually derived by the individual bodies in the context of their national conditions and in the language of the specific techniques or areas of testing.

The OECD Principles of GLP is an example of one such set of documentation developed internationally that applies to laboratories. Indeed, it is cited as such in the current (1990) edition of ISO/IEC Guide 25. Interestingly, the draft revision (1997) does not refer to the OECD Principles of GLP, emphasizing rather the relationship to ISO 9001/2 (*4*), (*5*). Over the years, requirements for accreditation and GLP have come closer together. The major differences are now in the practices used by monitoring authorities and accreditation bodies.

In countries where the accreditation body has responsibility for monitoring GLP compliance, that body applies the general criteria defined in ISO/IEC Guide 25 and superimposes the specific requirements of GLP for its evaluation of laboratories. There is no conflict between GLP and ISO/IEC Guide 25 and differences are largely only of emphasis. The accreditation body also applies the practices common to the monitoring authorities.

The differences in practice, referred to above, include the requirement of a full internal audit of every individual study carried out under GLP, whereas this is not normally the case under laboratory accreditation. Similarly, the concept of a single study director is virtually unknown in accreditation practice. In other respects, such as requirements for quality assurance programs and provision of suitable archives, there are strong similarities. In laboratory accreditation, the normal surveillance of accredited laboratories includes audits of laboratory data from a cross-section of the laboratory's work, but not the detailed study audits normally required by GLP monitoring authorities. Laboratory accreditation also appears to be more concerned than GLP with the sponsor (client) - laboratory relationship.

Verification of Compliance

Good Laboratory Practice. GLP compliance monitoring may take a number of forms. First, there are normally arrangements for ongoing routine surveillance of compliance. Second, GLP compliance monitoring procedures are often initiated as a consequence of a laboratory submitting data from studies, that are required to be conducted under GLP, to a government regulatory body. Special audits may also be sought by other countries or other domestic regulatory authorities. In general, therefore, GLP compliance monitoring is concerned with determining whether GLP compliance statements on certificates attached to studies are credible. Thus, GLP compliance monitoring activities focus substantially upon the audit of data of actual studies performed by a testing laboratory after the data have been submitted to and used by a regulatory authority.

It is the business of GLP inspectors to verify that a study was undertaken according to the plans and procedures decided upon in advance and that the complete study can be reconstructed based on the records kept. It is not their job to verify the scientific validity of the data or the accuracy of the conclusions of a study itself, only whether the data were derived and reported correctly, based on the study plan, Standard Operating Procedures (SOPs), etc. It is up to the regulatory authority, to which a study is submitted, to decide on the scientific accuracy of its conclusions.

Frequently, because of resource considerations, not all studies used by a regulatory authority can be audited for GLP compliance. Thus, the audit of a few select studies, largely on an *ad hoc* basis, becomes the means of assessing the general state of compliance of a laboratory with GLP for all submitted studies.

GLP compliance monitoring procedures are described in the Annexure to the 1989 OECD Council-Decision Recommendation on Compliance with Principles of Good Laboratory Practice [C(89)87(Final)] (*6*) and normally consist of biannual inspections of laboratories and audits by a GLP monitoring authority of studies that have been completed and submitted to regulatory authorities.

Laboratory Accreditation. Laboratory accreditation bodies also assess laboratories at regular intervals (surveillance). During these assessments the emphasis is on evaluating activities retrospectively through examination of reports and other records because the principal purpose of this process is to continue the accreditation and thus to provide assurance of ongoing technical competence and quality of test data to customers of the laboratory. Since laboratory accreditation is normally concerned with individual tests and measurements, rather than extensive long term studies, laboratory accreditation bodies do not usually have procedures to deal with auditing of single studies dealing with work which may have taken up to two years to complete.

When an accreditation body has responsibility for GLP monitoring there must, however, be provision for retrospective study audits upon request by regulatory authorities as part of the normal accreditation process.

Laboratory accreditation is directly concerned with the specific results of its assessments. When a laboratory is accredited for carrying out specific types of analytical tests, the accrediting body declares that the data were produced under a quality system designed to produce accurate and reliable data for those tests, and only those tests are covered by the endorsement of the accreditation body.

Laboratory accreditation, for both initial assessments and routine surveillance, uses a combination of peer review of technical competence through on-site assessment for compliance with published criteria, complemented where practicable through proficiency testing programs.

Auditors/Assessors

For international harmonization, there is concern that there is consistency of application of criteria and requirements and of interpretations made by individual auditors. The problem is well recognized and strenuous efforts are being made to internationalize the training of people who work in both areas.

An interesting difference between GLP monitoring authorities and accreditation bodies is that the former rely almost solely on professional staff ("inspectors") who work exclusively for the monitoring authority, while accreditation bodies use external expert assessors, in some instances supported by professional accreditation authority staff, for a small number of jobs each year.

Quality Systems Standards

It is occasionally suggested that compliance with a quality system standard (most often ISO 9002) provides sufficient confidence in a laboratory's test data (product). The ISO 9000 series of quality systems standards are, however, concerned solely with an organization's management system. Indeed, in the Introduction to ISO 9002 it is stated that the quality system requirements specified in the standard *"are complementary (not alternative) to the technical (product) specified requirements."* Compliance with ISO 9002 alone, therefore, makes no statement at all about the technical competence of a laboratory. In this the quality systems standards are similar to the OECD Principles of GLP.

In the introduction to the current (1990) edition of ISO/IEC Guide 25 it is stated that meeting the requirements of the Guide also meets the "relevant requirements" of the ISO 9000 series of standards (1987 edition). The scope of the draft revision of the Guide states that compliance with the Guide is compliance "with the requirements of ISO 9001 or ISO 9002" (1994). A cross-reference between the clauses of the Guide and those of ISO 9001/2 is given in Annex A of the Guide. The Guide separates management system requirements (section 4) from technical (competence) requirements (section 5).

Conclusion

ISO 9001/2 and the OECD Principles of GLP are management system standards, compliance with which gives confidence in the policies and procedures applied by a laboratory generating test data. An evaluation of the technical validity of the test data generated needs to be done separately. Laboratory accreditation against ISO/IEC Guide 25 gives confidence in the policies and procedures applied but also in the technical validity of the data generated.

While the requirements of GLP and Guide 25 are not identical, we submit that there is sufficient similarity so that, for (short term) analytical "studies", compliance with ISO/IEC Guide 25 could be accepted as compliance with GLP. In an addition, EAL Committee 3 has drafted a Guide setting out 22 "special interpretations" that could be applied to the various clauses in Guide 25 to meet GLP requirements. There are an additional 16 relatively minor requirements, relating to the QA statement on study reports, the responsibilities of the study director, archiving and report content, that would need to be incorporated into Guide 25 to demonstrate that compliance with the Guide 25 assured compliance with GLP. Compliance with ISO/IEC Guide 25, unlike compliance with the OECD Principles of GLP, also gives an assurance of the technical validity of the data generated.

Literature Cited

(1) *OECD Principles on Good Laboratory Practice (as revised in 1997),* Organization for Economic Cooperation and Development, Paris, France, Number 1, 1998.

(2) *ISO/IEC Guide 25:1990, General requirements for the competence of calibration and testing laboratories,* International Organization or Standardization/International Electrotechnical Commission, Geneva, Switzerland.

(3) *ISO/IEC Guide 58:1993, Calibration and testing laboratory accreditation systems - General requirements for operation and recognition,* International Organization for Standardization/International Electrotechnical Commission, Geneva, Switzerland.

(4) *ISO 9001:1994, Quality systems - Model for quality assurance in design, development, production, installation and servicing,* International Organization for Standardization, Geneva, Switzerland.

(5) *ISO 9002:1994, Quality systems - Model for quality assurance in production, installation and servicing,* International Organization for Standardization, Geneva, Switzerland.

(6) *Environment Monograph No 110, Guidance for GLP Monitoring Authorities, Revised Guides for Compliance Monitoring Procedures for Good Laboratory Practice, Council-Decision Recommendation on Compliance with the Principles of GLP [*

Chapter 5

The OECD GLP Principles and ISO/IEC Guide 25: Schisms and Bridges

Theo Helder

GLP Monitoring Unit, Veterinary Public Health Inspectorate, Ministry of Health, Welfare and Sport, P.O. Box 5406, 2280 HK Rijswijk, the Netherlands

Both the Good Laboratory Practice Principles and the ISO/IEC Guide 25 are quality systems to be applied to research. Because of this the systems are quite comparable and have much in common. Basic differences between the criteria, like the independent QA function, the study director concept and archiving requirements, arise from the distinct scopes of GLP and ISO 25. The application of GLP is legally required for research in the regulatory area, whereas ISO 25 serves almost exclusively the private market. Therefore, only a few laboratories need to implement both quality systems. As a result of the distinction of scope, the monitoring of GLP compliance and ISO-type laboratory accreditation serve different purposes and are performed in quite different ways. GLP compliance has to be monitored by government inspectors with investigational powers and authority, and ISO-conformity assessment is conducted by private bodies. Some monitoring authorities issue "Statements of Compliance" after satisfying inspections, which is essentially different from the formal accreditation performed by accreditation bodies. Theoretically, it is possible to perform laboratory "accreditations" for OECD GLP compliance; however, the task of performing directed inspections and study audits is and must remain the responsibility of government bodies. Therefore, the practicality of contracting out part of the monitoring programmes to private bodies is implausible.

The implementation of rules or principles of Good Laboratory Practice in the conduct of non-clinical and safety studies became a legal requirement in almost all of the OECD countries in the late seventies until the mid eighties. Monitoring programmes were subsequently set up by the OECD Member Countries. Not long after, industry became aware of the advantages of also having a quality system for the non-regulated area, as a result of which the ISO 9000 series standard and, more specifically, for laboratories, ISO Guide 25 and its equivalent European standard, EN 45001, were developed. Governmental or private organisations took up the task to accredit the laboratories implementing these quality systems. Because of the similarities between the requirements of both the Principles of GLP and the ISO/IEC Guide 25, and the purported burden of multiple inspections, a movement has developed to bring together the ISO 25 and GLP quality systems. Although it is clear that on paper both quality systems are much alike, the differences between their origins, scope, purpose and legal position make a merger virtually impossible.

Implementation and Monitoring

When comparing the GLP system with the ISO or EN system, the very first thing to be done is to discriminate between the rules or principles to be applied in the laboratory or test facility and the surveillance of their application by third parties or authorities. This is very much neglected by those in favour of merging the two quality systems. It is of major importance to separate implementation and surveillance in order to get a clear overall picture. The "GLP-system" is based on the Council Decision of the OECD Council on Mutual Acceptance of Data [C(81)30(Final)] and Council Decision C(89)87(Final). These decisions stipulate that data of safety research produced in one Member Country will be accepted in any other Member Country, if the Principles are applied and an appropriate monitoring is in place. The principles to be applied are laid down as Annex II to the C(81)30(Final) Decision (*1*), whereas the guide for monitoring procedures and the guidance for laboratory inspections and study audits are to be found as the Annexes I and II of the Council Decision C(89)87(Final) (*2, 3*).

A similar partition is to be found in the area of laboratory accreditation. The standard, EN 45001, that is equivalent to ISO/IEC Guide 25, sets out the requirements that laboratories have to meet; EN 45002 and 45003 set out the rules for the assessment of testing laboratories and for laboratory accreditation bodies, respectively.

Comparison of Requirements

Several efforts to compare GLP and ISO/IEC Guide 25, or its equivalent EN 45001, have been made. From these, it is quite obvious that GLP and ISO/IEC 25 have different starting points and different endpoints. Whereas ISO 25 requirements are fully in line with the laboratory accreditation requirements, i.e. aimed to demonstrate the competence to carry out a specific task and, therefore, is test method oriented, GLP is study oriented and designed to fulfill governmental requirements to demonstrate that a study is fully reconstructible from the study plan, SOPs, raw data and final report. This means that GLP has stringent requirements related to a study plan, study director, final report, study director's GLP statement, and the QA statement that cannot be found in ISO 25. Also there is a requirement for an independent QA programme, which is considered a very strong requirement for GLP, very much unlike the ISO's self assessment scheme. Furthermore, GLP requires complete recording of all events and has strict archiving rules, which also cannot be found in ISO 25. To be compliant, a laboratory, therefore, should apply the GLP principles over the full width of its activities. For each study, it should be decided if GLP rightfully can be claimed, dependent on whether or not there is full adherence to the GLP Principles. It should not be forgotten here that adherence to the GLP Principles is a legal requirement for safety studies.

ISO 25 laboratory accreditation is aiming at the assessment of the competence to perform specified tests. These tests are of a repetitive nature unlike GLP studies. The laboratory is free to decide which and how many types of tests are to be assessed. This makes it possible that a laboratory is accredited for only a fraction of the types of tests conducted.

Because of these differences, ISO 25-type testing might only be used instead of GLP where it concerns repetitive testing activities like urinalysis, blood chemistry, etc., as part of sizable studies. Also, it might be feasible for very short term studies as physical-chemical studies, Ames testing, etc., as long as all GLP requirements are fulfilled.

The OECD GLP Panel, consisting of the heads of all monitoring units of the OECD Member Countries, has considered also the use of laboratory accreditation with reference to GLP monitoring and has issued a position paper (*4*), stating that "data, generated solely under ISO/IEC Guide 25 or equivalent standards is unlikely to be accepted by regulatory authorities.....".

Simultaneous Implementation of GLP and ISO 25

The fact that ISO 25 and GLP are not equivalent does not mean that simultaneous implementation of both systems is impossible. A EUROLAB-EURACHEM Working Group has published a helpful guide to simultaneous implementation (*5*). Almost all OECD Member Countries' monitoring units have reported successful simultaneous implementation of both GLP and ISO 25 in a number of test facilities. However, the number is quite limited when compared with the total number of laboratories in the

Table I. Number of Test Facilities in the European Union Implementing GLP and GLP Plus EN 45001 (1996)

Country	*GLP*	*GLP + EN 45001*
Belgium	26	2
Denmark	17	2
Finland	8	0
France	94	?
Germany	170	10
Greece	6	?
Ireland	6	0
Italy	25	?
Netherlands	50	3
Portugal	2	0
Spain	6	?
Sweden	30	?
United Kingdom	160	8

respective Member Countries in the EU. Table 1shows the results of a survey among the EU Member States.

Because of these low numbers, it was concluded by the European Commission, in a meeting of the EU Working Party on GLP, that the search for possibilities of merging GLP and EN 45001 would be given low priority.

Scope of GLP and EN 45001

Internationally, the scope of GLP has been well defined. According to several European directives, to be incorporated in EU Member-States' legislations, all preclinical and safety studies on industrial chemicals, pharmaceuticals, veterinary drugs, animal feed additives, pesticides, and cosmetics have to be carried out in compliance with the OECD Principles of GLP. Basically, the same requirements exist in the other OECD Member Countries. Some countries have even a more extended scope; e.g., the USA requires application of GLP for medical devices and implants. The European Directives on medical devices and active implants just require the application of a quality system; both GLP and EN 45001 are acceptable. The scope for ISO/IEC 25 and EN 45001 is not very well defined, since these standards are almost exclusively applied in the non-regulated area. These standards are used mainly in the fields of analytical chemistry, physical chemistry, physics and clinical (pathological) chemistry, where there is a link with the regulated GLP area. In the European Union, there is only one legal requirement to apply EN 45001, i.e., in Directive 93/99/EEC on the supervision of foodstuffs. The Directive requires the application of EN 45001 plus OECD GLP Principles 2 (on QA) and 7 (on SOPs), which is quite incomprehensible since EN 45001 has definite paragraphs on SOPs.

Monitoring and Surveillance

According to Annex I to the OECD Council Decision, monitoring of GLP compliance has to be carried out by or on behalf of the governments of the OECD Member countries. This is understandable since governments are responsible for the admission of chemicals on their markets. Most governments have charged governmental or government-related bodies with this task. In some countries these bodies also have been charged with the task of laboratory accreditation, like in Australia (NATA), Ireland (ICLAB), New Zealand (Telarc) and Norway (NMS). In some other countries accreditation bodies perform GLP inspections only where it concerns industrial chemicals. This is the case in Denmark (DANAK), France (Cofrac), Portugal (IPQ) and Sweden (Swedac). Of the total of 32 monitoring authorities, 2 are private, 3 are government-related, and 27 are fully governmental. All these monitoring authorities perform their inspections according the relevant OECD Guidance (*3*).

The purpose of GLP monitoring is to determine that a study is reliable and reconstructible from the study plan and raw data and that it is reported correctly. The purpose of laboratory accreditation is to assess if the laboratory is competent to perform the tests as specified in their documentation. Therefore, other than in laboratory accreditation, GLP monitoring authorities have 2 major instruments. One is the typical test facility inspection, always including the limited auditing of studies. The other tool is the in-depth study audit requested by regulatory authorities (directed study audit). Since the application of GLP is legally mandatory, the inspectors performing such inspections and study audits should have investigational powers, for instance to enter the test facility's premises and inspect confidential material. In many legal systems it is not possible to grant such powers to private (laboratory accreditation) bodies.

Many monitoring authorities inspect the test facilities in their programmes on a regular basis, e.g. once in two years. Apart from the study audits, these inspections might be considered as "accreditations", since inspectors are assessing the general performance and the competence of the test facility. Many monitoring authorities even issue a "Statement or Endorsement of Compliance" when a test facility is judged to operate in compliance at the time of the inspection. This is not equivalent to certification as in the accreditation environment where it means an acknowledgement of competence; however, there is a tendency to use it in that way. It is thinkable to contract out this part of the monitoring programme to third parties, for instance accreditation bodies. These bodies then must follow the OECD Guide and Guidance. However, the task of performing study audits, being a regulatory task, must remain with the governmental authorities. For practical reasons it is preferable to leave both instruments in one hand: the governmental or government-related inspectorate.

Conclusions

1. ISO/IEC Guide 25 and equivalent standards are not equivalent to the GLP Principles.
2. Simultaneous implementation of GLP and ISO/IEC Guide 25 is quite possible.
3. Adherence to the GLP Principles is a legal requirement and should be monitored by governmental authorities.
4. OECD procedural guides Environmental Monographs Nos. 110 and 111 must be followed for GLP monitoring.
5. Monitoring authorities must have legal investigational powers.
6. Accreditation to OECD GLP is possible, but can only be part of GLP compliance monitoring.

Literature Cited

1. The OECD Principles of Good Laboratory Practice, Environment Monograph No. 45, OECD, Paris, 1992.
2. Revised Guide for Compliance Monitoring Procedures for Good Laboratory Practice, Environment Monograph No. 110, OECD, Paris, 1995.
3. Revised Guidance for the Conduct of Test Facility Inspections and Study Audits, Environment Monograph No. 111, OECD, Paris, 1995.
4. The Use of Laboratory Accreditation with Reference to GLP Compliance Monitoring: Position of the OECD Panel on Good Laboratory Practice, OECD, Paris, 1994.
5. Quality Assurance According to EN 45001 and OECD GLP. A guide to simultaneous implementation. The Joint EUROLAB-EURACHEM Working Group, EUROLAB T-QA WG2/EURACHEM, Netherlands WG3.

Chapter 6

The Impact of Good Laboratory Practice Accreditation Programs on Laboratory Operation

Patricia Royal

Quality Systems Consultants, Inc., 80 Main Street, Plympton, MA 02367

Discussions by Good Laboratory Practice (GLP) regulators, sponsor organizations, and laboratories on the advantages and disadvantages of developing a Laboratory Accreditation program for GLP testing laboratories are ongoing. These discussions are a result of laboratory testing programs expanding their markets into international territories, and regulators and sponsors seeking verification of GLP compliance status in distant locations.

The impact of developing a GLP Laboratory Accreditation program is still to be determined. Regulators, sponsors, and testing laboratories each have legitimate concerns. Regulators are concerned about maintaining control over the monitoring process. Multinational registrants of products must meet conflicting international testing requirements. Laboratories need access to satisfactory documentation attesting to their GLP compliance status.

Two quality standards exist today that affect laboratory operations. The first is the traditional GLP program. It is government-operated and is implemented by each participating country. The second is the International Organization Standards (ISO) Guide 25. While these two quality standards have similar components, their purpose, the user and implementation differ. The scope of GLP is in pre-clinical testing programs to determine risk by regulators, while the scope of ISO is broader, and is often referenced in product trade agreements. Before any GLP Accreditation program could successfully operate, recognition of the similarities and alignment of the differences between these two programs is needed. Optimally, a new standard could be developed and accepted by all parties, where regulators maintained authority, but managed by a third party.

Two separate international quality programs exist to evaluate laboratory operations. The first is the government operated Organization for Economic Cooperation and Development (OECD) Good Laboratory Practice (GLP) program (*1*), and the second is a privately operated program under the International Organization of Standards (ISO) called ISO/IEC Guide 25 (*2*). When the OECD GLP Principles were first issued in 1981, one goal was to reduce non-tariff barriers and open the large U.S. market to international registrants by implementing an international standard for data quality. ISO programs have developed with similar goals. The first ISO standards affecting data quality were the ISO 9000 series for Management of Quality Systems. The ISO/IEC Guide 25 is the management systems standard applied to laboratory operations.

The registration of agricultural products has become a global process. Currently, there are greater registration activities outside of the United States, particularly in Europe, South Africa, and Latin America, than in the United States. Governments in many of these countries have implemented limited or no GLP test monitoring programs, while others, including Canada and South Africa are considering alternative programs that include third party monitoring.

Debates on the pros and cons of developing a government recognized third party laboratory accreditation program for GLP testing programs are important. The outcome could impact the direction of research and development programs of regulated products for many years. There is a real need for better understanding of what accreditation is, and what it is not; the origin of these debates; and how a GLP Accreditation program would affect regulated laboratory operations as we know them today.

OECD GLP Principles and ISO Standards

Both the OECD GLP Principles and the ISO programs are international quality standards developed to promote uniformity of data quality. While these programs have common goals, their purpose and implementation are different. The GLP program is a regulatory process developed to ensure the quality of data used by regulators to support registration of products and to determine risks to health and environmental safety. The ISO process, on the other hand, is primarily a privately operated voluntary program used to set standards to promote trade of products between nations. Commerce issues relate to Trade Agreements, the most significant being the North American Free Trade Act (NAFTA) (*3*) and the General Agreement on Tariff and Trade (GATT) (*4*). Because the ISO is more oriented to trade issues, it has taken a more prominent role in these Trade Agreements than has GLP, which is implemented by each country and has country-specific requirements. Thus, there is little mutual recognition in the effectiveness of these two programs.

Criteria for GLP Program Changes

GLP is a regulatory process. If a GLP Accreditation program were to be developed, it would need sanction by international authorities. Three key restrictions and criteria would have to be met in order to accommodate the needs of regulators, registrants and laboratories alike.

First and foremost, the program would need to be based on components of OECD GLP. In order to gain international acceptance, the program would need to meet criteria outlined in the OECD Guidance Documents, which specify government interaction and control of the monitoring process. In particular, OECD has established significant restrictions on laboratory accreditation programs (*5*)(*6*). In 1994, the OECD GLP Panel issued an Opinion Paper on laboratory accreditation programs. There it stated that accreditation programs based solely on ISO 9000 and Guide 25 would not by themselves be sufficient for registration purposes. These conclusions are based on the premise that because ISO programs are privately managed and process-oriented, and, as such, lack government control, do not address study-specific events, and data quality (*7*). Therefore, a new standard based on GLP requirements would need to be developed.

Secondly, economic feasibility would be an essential component to the success of any GLP Accreditation program. A poorly designed accreditation program might elevate costs to governments, but it could put an excessive financial burden on registrants and private contract facilities, which could undermine the success of any program. Therefore, precautions would be needed to ensure that the program is cost-effective for all participants.

Thirdly, any accreditation program must be compatible with existing systems, namely the GLPs as they function today, and meet international sanctions. GLP is a governmental regulatory process, and therefore, not easily changed. However, there is nothing in either the OECD or GLP programs in the U.S. to prevent the development of a GLP Accreditation program. OECD GLP requires that any accreditation program has government oversight and sanction and be based on GLP requirements.

What is Accreditation?

Accreditation is an evaluation process by an appointed authoritative body (in this case, of a laboratory) which is given recognition when found competent to carry out specific tasks (in this case, to GLP standards) (*8*). This evaluation would be a technical as well as a quality standard evaluation. Therefore, a laboratory could be accredited for GLP competency and technical discipline in environmental sciences and/or mammalian toxicology. Because the accreditation process is systems-oriented rather than study-oriented, it de-emphasizes individual roles, such as the Study Director, and emphasizes the process of producing quality data and maintaining control over the process of collecting and reporting data.

Accreditation does not assure compliance of a program or of a particular study, but rather evaluates the competency of a laboratory to conduct a study in compliance with GLP. In the U.S., it must be remembered that the Enforcement Policy is based on the Compliance Statement, which is study-specific (*9*).

A big difference between the two programs is that laboratories are active partners in the evaluation process and request accreditation and inspections by an approved accrediting organization. Therefore, if a laboratory elected to participate, the accreditation organizations would assure the adequate frequency of inspections and evaluation against an accepted international standard. Adequate frequency is considered to be every two to three years, with yearly surveillance audits between full accreditation evaluations (*10*)(*11*). This is a significant benefit over the current program.

Cost would be a big factor in the successful development of a GLP Laboratory Accreditation Program. It has been difficult to estimate the cost of the current U.S. Program; both to industry and the US government. EPA currently estimates approximately 1500 - 2000 facilities operate under the FIFRA program. Their current budget does not allow for inspecting facilities every other year.

How Accreditation Differs from Current GLP Programs

Both GLP and ISO programs have standards or guidance documents, which describe the monitoring authority, its implementation and the inspection process. A difference is that accreditation programs generally pre-qualify laboratories. Therefore, the competency of a laboratory to conduct studies according to GLP requirements could already be evaluated before conducting any study.

Accreditation evaluates a process, such as document control, rather than the compliance of a specific study to GLPs. Study-specific GLP audits could be a part of the process, and study-specific audits are often a part of the ISO Guide 25 evaluation process. The concept of process-oriented audits vs. study-specific audits is a significant difference between the current U.S. EPA and FDA programs and traditional accreditation schemes.

Accreditation is financially self-sustaining, because the participants finance the program. The laboratory would pay an Accrediting Organization to become accredited. Some see this as a weakness of the accreditation process because both the buyer and the seller are outside of government operations. This could lead to a potential conflict of interest. This potential conflict of interest is a significant concern of regulators. Regulators have speculated that if they lost their management control to the private sector, selection of Accrediting Organizations by the laboratory would be based on cost and could weaken the current process. However, large international accrediting associations (International Laboratory Accreditation Council (ILAC)) monitor Accrediting Organizations to assure their quality. Likewise, the reputation and market competitiveness between accrediting organizations and government oversight would be built into the system and would minimize this concern.

Accreditation programs are generally privately operated, either with or without government sanction. However, government sanction would be an absolute requirement of any GLP Accreditation program.

Evaluation Process

The criteria for GLP Accreditation are government sanctioned and based on GLP standards, not on the ISO Guide 25 Standard. Beyond that, the evaluation process can be described as laboratory interactive and process oriented.

Once a standard is developed and accepted by regulatory authorities, accrediting organizations would develop accreditation programs to be approved by EPA or other regulatory bodies. Once approved, laboratories would contract with these accrediting organizations to accredit their facilities. The process begins by submitting documents that describe laboratory activities, such as study types, SOP index, key job descriptions and resumes, etc., to the accrediting organization for review. Generally, the next step is a pre-inspection visit or meeting. This is not an inspection as such, but rather a meeting to lay the groundwork and identify gaps to promote successful accreditation by the laboratory. A mutually acceptable inspection date is then set in advance.

The inspection generally evaluates systems from a horizontal rather than vertical (study-specific) point of view. Therefore, it evaluates the function of the management, QAU, SOPs, maintenance and calibration of equipment, record keeping and data storage.

Comparison Between ISO Guide 25 and GLP Standards

When we speak of Laboratory Accreditation, we think of ISO Guide 25, which does have components comparable to GLP, but also, it has significant differences. ISO Guide 25 is an international standard used to evaluate laboratories. While acknowledging that modifications to ISO Guide 25 are needed to meet GLP standards, these two standards do have many similar components. Thus, the ISO Guide 25 will be used as a benchmark in this paper to compare it with requirements found in the OECD GLP Revised Standard. Comparison of sections is given in Tables I and II.

Table I. Comparison of GLP and ISO Standards

GLP	ISO Guide 25
	• Process Oriented(#5.1)
• Study Oriented(#8.1)	• Organization & Management(#4)
• Organization & Management(#1)	• Quality Standards & QA/QC(#5.1& #5.2)
• Quality Standards & QA(#2.2)	
• Test Item labeling, storage & handling(#6.1)	• Test Item labeling, storage & handling(#11.1)
• Personnel Training(#1.4)	• Personnel Training(#6.1)

Table II. Comparison of GLP and ISO Standards

GLP	ISO
• SOPs for Procedures & Calibration (#7)	• SOPs & Test Methods(#10)
	• Archiving(#12)
• Archiving (#3.4)	• Final Product Control(#5.3 & #5.6)
• Final Report Review(#2.2)	• Report(#13)
• Report (#9.1& #9.2)	

The ISO Guide 25 requires a Quality Manual. The contents of the Quality Manual and related documents are given in section 5.2. Many requirements found in the GLPs are specified in this section of ISO Guide 25. Also included are requirements for Management commitment that are essential to develop a quality program, including the appointment of a Quality Manager. The Quality Manager is responsible for implementing the quality program including the Quality Assurance Program. Duties of the Quality Manager, while not as prescriptive in ISO Guide 25 as in the GLP Standard, are not incompatible with those found in the GLPs. Compatibility with the GLP

Standard would require the addition of study specific inspections and data audits, along with maintenance of the Master Schedule and Protocols. The Quality Manual also requires "procedures for departing from standard methods" (protocol deviations), "procedures for maintenance and calibration of equipment" and procedures for audits, along with other requirements similar to GLP. The difference is that in ISO Guide 25, procedures used by the laboratory must be referenced or written and organized in a Quality Manual and are process-oriented.

Sections 4 and 5 of ISO Guide 25 require organization (charts), management, and the quality assurance review process. Section 6 specifies staff training requirements that are essentially identical to those found in the GLPs. It states that for any assigned task, "Training records/job descriptions must be kept current...the lab shall have sufficient personnel having education, training, technical knowledge and experience." Section 7 of ISO Guide 25 specifies that environmental conditions must be such as not to interfere with the study, and that adequate space must be available. Section 7.3 states that "the laboratory shall provide facilities for the effective monitoring, control and recording of environmental conditions..." Section 7.4 states that "there shall be effective separation between neighboring areas where activities are incompatible." Sections 8 and 9 of ISO Guide 25 specify equipment maintenance and calibration requirements. These requirements are more specific in ISO Guide 25 than those found in the GLPs. Section 9.1 of ISO Guide 25 states that laboratories shall have established programs for maintenance and calibration verification. Section 9.2 states that wherever possible, validation should be designed to traceable standards. Section 9.4 requires that Reference Standards shall be used only for calibration. Section 12 of ISO Guide 25 gives record-keeping requirements. These requirements pertain to data retention only, and do not address the real time data collection requirement found in the GLPs.

The GLPs are study-oriented, while the ISO Guide 25 is process-oriented. This is a significant difference between the two systems, but not insurmountable as vertical, study-specific audits are often conducted. Both standards have similar requirements for defining organization and management responsibilities.

Quality Standards as defined for the Quality Assurance Function differ somewhat; the GLP program being more specific, but they are compatible with modification to include study specific inspections. Requirements for test item labeling, storage and handling are essentially identical, as are requirements for personnel training, records and job descriptions.

Record-keeping requirements in ISO Guide 25 (section 12) are less specific that those found in GLP and would need modification to meet GLP requirements of real time data documentation and collection. Section 7 of the OECD GLP program provides requirements for SOPs. These requirements include SOPs for routine procedures and calibration of instruments. Because GLPs are oriented to specific studies, test method requirements are specified in Protocols. The ISO standard is process-oriented, thus, test methods are more standardized than in GLPs and often follow published methods. Section 10 of ISO Guide 25 gives requirements for SOPs and the use of standardized test methods. They take the place of protocols. Standardized test methods are those such as ASTM, EPA or OECD. GLPs, on the other hand, require that test methods and procedures be described in a protocol and signed by the responsible Study Director. There is no requirement for a QA Statement or a Compliance Statement in the ISO report section because this Standard has not been used as a regulatory process. These are significant differences between GLP and ISO. Section 3.4 of the OECD GLP gives requirements for archiving data at the conclusion of a study. Similar requirements, but fewer specifics are found in ISO Guide 25, Section 12. Both standards require final "product" or data review, although not as prescriptive in ISO as in GLP. They both call for essentially identical reporting requirements.

When the two standards are compared, similar components are evident in organization, management, SOPs, maintenance and calibration of equipment, and handling and labeling test items; but there are also significant differences. A summary of similarities and differences is given in Table III.

Table III. Significant Similarities and Differences Between GLP and ISO Guide 25 Programs

Similarities	Differences
• Organization • Management • Quality Assurance Officer • Personnel Training and Records • SOPs • Maintenance and Calibration of Equipment / Environmental Accommodation • Labeling, handling and storage of test item • Final Report and Product Assessment • Archiving	• Process oriented vs. study oriented • Privately operated vs. Government operated • Function of QAU • De-emphasized role of individual such as Study Director • Use of standardized test vs. protocols • Requirements for Biological Components • Real-time record-keeping • QAU and Compliance Statements

Differences include process-oriented audits vs. study-oriented audits, which de-emphasize the role of individuals, such as the Study Director. Another difference is that ISO generally operates independently of the regulatory process, while GLP is a regulatory government operated process. Yet another significant difference is that ISO Guide 25 uses standardized tests, while GLP Standards require study-specific protocols, which are research-oriented and subject to change. There are no specific biological component requirements in the ISO Guide 25 Standard. This is a significant drawback when one realizes that a large segment of GLP testing programs pertain to biological testing schemes to determine risk to health. Real time record-keeping requirements also vary considerably between the two standards.

The real differences between accreditation and GLP, however, are not so much in the standards themselves as in the purpose, the use, and the implementation of the two programs. ISO Guide 25 is primarily oriented to trade and standardization of registered products, while GLP is oriented to government control of quality during the development of data used to evaluate the risk of products that are to be registered.

Financial Considerations

There are concerns by laboratories and sponsors that if an accreditation program were developed, it would be an added cost layered onto the current expense already paid by sponsors. A legitimate request by sponsors is that any added cost must have added value or provide a trade-off with current requirements and liability, such as those specified in the U.S. EPA Enforcement Policy. It is difficult to estimate actual costs of the current program managed by EPA, costs to industry, and costs of developing new programs, whether it be accreditation or some other option. It is equally hard to predict which system the international market will favor. Therefore, it is difficult for the industry to know where to invest its time and money.

EPA has reported an Office of Enforcement Compliance Assurance (OECA) travel budget of $100,000/year (*11*). This budget is used to inspect approximately 100 - 150 facilities per year. There are a reported 1500 - 2000 facilities operating under GLP/Federal Insecticides, Fungicides, and Rodenticides Act programs in the United States. Adequate financing to inspect all facilities at an acceptable frequency (every 2 to 3 years) is a shortcoming of the current U.S. program. The current budget allows for approximately 200 - 300 biannual inspections. If extended to a minimum of 3 years, the number of 350 - 400 inspections still falls short of optimal inspection frequency.

If U.S. laboratories are to compete with international laboratories, the U.S. program must be augmented with additional funds, outside programs, or both to ensure that the frequency of inspections meets the international demand of regulators and registrants. An accreditation program could act as a voluntary program to augment this process, or it could act as a mandatory, stand-alone program. It has been estimated that a GLP Accreditation program would cost the laboratory an estimated $6,000 for the inspection and an additional $1,000 for administrative costs, thus totaling $7,000 for each participating facility for an accreditation lasting 2 years (*11*).

It cannot be forgotten that industry and the U.S. government have invested heavily in the currently operating GLP program, and that the U.S. EPA and FDA programs are essentially harmonized with those of the OECD, making a relatively seamless international GLP program. This harmonization is a real benefit when considering that the program is time-tested and that many laboratories conduct studies for all three regulatory bodies. Current industry yearly investment in the GLP program under FIFRA has been estimated to be at least $30,000,000 (*11*). The U.S. regulated Industry and U.S. EPA are in agreement that change cannot come arbitrarily and must provide substantial value and incentives to be accepted.

Advantages and Disadvantages of Accreditation

The development of a U.S. GLP Accreditation program would have significant advantages and disadvantages over the currently operating U.S. EPA program. Advantages and disadvantages are presented in Table IV. The first most obvious advantage is that the program is financially self-sustaining. The laboratory pays a fee to be inspected and accredited by an accrediting organization. The disadvantage is the potential for a conflict of interest. At present, we have only an estimate of the current cost to EPA and the cost to accredit a facility.

Table IV. Advantages and Disadvantages to GLP Accreditation

Advantages	Disadvantages
• Financially Self Sustaining	• Cost Hard to Predict
• Integration of Trade and Regulatory Issues	• Does Not Assure Compliance
• Assures Adequate Number of Inspections	• Questionable International Government Acceptance
• Available Trained Inspectors	• Potential Conflict of Interest

The next advantage is the integration of trade and regulatory issues. This is a real incentive in developing an accreditation scheme. If initiated, it would be imperative to ensure that international governments accept a 3rd party program. Ensuring adequate frequency of inspections of all facilities would be a significant benefit for accreditation. However, it must be remembered that while accreditation would assure an adequate number of inspections, it would not assure study compliance. Regulator concerns focus on ensuring their adequate control over the monitoring process. Regulators use data from GLP studies to determine risk to human health and the environment. It is a regulatory process under their jurisdiction. Their concern is over deputizing the monitoring process to the private sector and the potential weakening of the current system.

Sponsors are concerned both about their liability if an accreditation program were developed and the cost to implement such a program. In order to be successful, compromises would be needed, including modification to the Enforcement Policy. It is accepted that if a GLP Accreditation program were developed, it would have to meet certain criteria based on internationally agreed upon standards. Any program would have to be based on the OECD GLP program, and meet the Mutual Acceptance of Data Agreement (MAD) (*12*). It would also need to be conducted according to the OECD Guidance Documents and meet the OECD criteria for Accreditation (*13*) (*14*).

Compatible components must be identified and differences in approach must be integrated, including vertical vs. horizontal assessments and government sanctioned 3rd party management. Accomplishing these tasks would do much to integrate country specific regulatory requirements and those requirements specified in trade agreements and reduce the tension between these two operations.

Summary

There is a growing interest in developing a GLP Standard for Laboratory Accreditation or incorporating GLPs into existing standards. Many similar components affecting laboratory operations could be built into a GLP Laboratory Accreditation program.

These include management, organization, SOPs, personnel, QA functions, study plans, study conduct, and archival of data.

Significant differences do exist, however, between these two programs, and these must be resolved before an accreditation program could be developed. Significant differences in implementation include process-oriented vs. study-oriented audits and privately operated vs. government-operated programs. Differences in standard requirements include study management/control vs. process management, data collection, biological components, and record keeping including archiving.

Major factors to the successful implementation of such a program include cost containment, compatibility with OECD GLP programs, frequency of inspections, and mutual recognition of data and monitoring programs.

Developing a GLP Accreditation program will be a conscientious decision made after much thought on the current U.S. and international programs. EPA's evaluation of the ELAB GLP Subcommittee Report, along with international needs, will impact any decision to develop such a program.

Acknowledgments

I would like to thank Deborah Muise for her help in the preparation of this manuscript.

Literature Cited

1. OECD Principles of Good Laboratory Practices; (C(81)30(Final) Annex 2), 1998.
2. ISO/IEC Guide 25, Geneva, Switzerland, 1990.
3. North American Free Trade Act, House Document, pp. 103-159, Vol. 1, 11/4/1993.
4. General Agreement on Tarriff and Trade, Uruguay Round, April 15, 1994 (update 1996).
5. OECD: **1994**; "Issues Arising from the Panel on GLP"; Environmental Policy Committee, Chemical Group and Management Committee.
6. OECD: **1993**; "Position Paper on GLP Compliance Monitoring and Laboratory Accreditation".
7. Royal, P., **1994**; *Quality Assurance, Good Practices, Regulations and Law,* "Harmonization of Good Laboratory Practices Requirements and Laboratory Accreditation Programs", Vol. 2, No. 3.
8. ISO Guide 2, Geneva, Switzerland, 1986.
9. U.S. EPA: **1991**; FIFRA GLP Standard Enforcement Response Policy.
10. A2LA, 1993 General Requirements for Accreditation.
11. U.S.EPA: **1997**, ELAB GLP Subcommittee Report.
12. OECD: **1981** Mutual Acceptance of Data in the Assessment of Chemicals [C(81)30(Final)].
13. OECD: **1995** Environmental Monogram #110; "Guidance for GLP Monitoring Authorities; Revised Guide for Compliance Monitoring Procedures for Good Laboratory Practices".
14. OECD: **1995** Environmental Monogram #111; "Guidance for GLP Monitoring Authorities; Revised Guidance for the Conduct of Laboratory Inspections and Study Audits").

Chapter 7

Third Party Laboratory Accreditation: A2LA—A Case Study

Roxanne M. Robinson

The American Association for Laboratory Accreditation, A2LA, 656 Quince Orchard Road, #620, Gaithersburg, MD 20878

Laboratory Accreditation within the United States is gaining favor. Confidence in test data is paramount to product acceptance. Users are looking for assurance of high quality products and the means to evaluate suppliers without incurring the costs associated with auditing each supplier. Reliance on third party accreditation to perform this function is an attractive option. The use of ISO/IEC Guide 25 as the accreditation criteria is also considered valuable for international acceptance of test data.

The American Association for Laboratory Accreditation (A2LA) is a third party accreditor offering a broad based program of testing laboratory accreditation using ISO Guide 25 as the accreditation criteria. This chapter describes the A2LA organization and its domestic and international activities. The requirements of Guide 25 will be discussed, as well as the accreditation process, from application to granting of accreditation.

ISO/IEC Guide 2 "General terms and their definitions concerning standardization and related activities" (*1*) defines "accreditation" as the procedure by which an authoritative body gives formal recognition that a body or person is competent to carry out specific tasks.

A2LA is a nonprofit, professional association with individual, institutional, and organizational members. Membership is open to anyone. The Association's goal is to provide a comprehensive national laboratory accreditation system which establishes widespread recognition of the competence of accredited laboratories. Elimination of the unnecessary multiple assessment of laboratories is also a goal.

A2LA has been accrediting laboratories since 1980, using the international standard ISO/IEC Guide 25 "General Requirements for the Competence of Calibration and Testing Laboratories" (*2*) as its criteria for accreditation. This standard not only requires a quality system and manual in the laboratory but also requires that the laboratory be found competent to perform specific tests and types of tests. In 1990, Guide 25 was revised, considering the content of ISO 9002 (Quality Systems - - Model for Quality Assurance in Production, Installation, and Servicing) (*3*) and is presently undergoing another revision. A2LA's operations are designed to meet the requirements of ISO/IEC Guide 58, "Calibration and Testing Laboratory Accreditation Systems -- General Requirements for Operation and Recognition" (*4*).

A2LA currently accredits laboratories in ten different fields of testing, including Acoustics & Vibrations, Biological, Chemical, Construction Materials, Environmental, Geotechnical, Mechanical, Nondestructive and Thermal. The Association also offers accreditation in the Calibration field. A2LA accredited laboratories currently number 1,019. Funding comes from membership dues, fees for services, and training courses.

Organization

A2LA is supported by volunteering groups which provide administrative and technical guidance and support to the A2LA staff and assessors.

Board of Directors. A2LA is governed by a 22 member Board of Directors. Board membership represents interests in industry, labor, laboratories, government and the professions. Two Councils, the Accreditation Council and Criteria Council, report to the Board of Directors. Officers of the Board include the Chairman, Vice Chairman, Secretary, Treasurer, Accreditation Council Chairman and Criteria Council Chairman. Two from A2LA Staff, the President and the Vice President, are also Officers of the Board. The Board is responsible for the management and administration of the Association. The Board meets three times per year and Board members can serve for a maximum of nine years.

Accreditation Council. Members or the Accreditation Council come from laboratories, industry and government. All are quality minded and have technical backgrounds in the fields of testing or calibration in which the Association accredits. The Accreditation Council makes decisions on granting, denying, or withdrawing accreditation based on the written evaluations provided by the A2LA assessors and Staff. All voting is done by letter ballot and a panel of three from the 42 member Accreditation Council is chosen to review and vote on each accreditation action. Panels are formed to match the technical discipline of the applicant laboratory. Appeals to a panel vote are decided by a vote of the full Accreditation Council. Appeals beyond the Accreditation Council are handled by the Board of Directors. Their decision is final.

Criteria Council. There are fourteen members of the Criteria Council and they are responsible for reviewing and approving criteria documents developed or revised by the Association. The Criteria Council is comprised of members with technical background sufficient to knowledgeably review technical requirements or criteria which make up A2LA's accreditation programs.

Assessors. A2LA contracts assessors on an as-needed basis to perform the laboratory assessments. A2LA assessors are drawn from the ranks of the recently retired, consultants, industry, academia, government agencies, and from the laboratory community. At minimum, each candidate must have at least five years of direct laboratory experience in order to qualify as an assessor. Auditing experience is a plus. They must be technically very knowledgeable in the field(s) in which they assess. They must also be knowledgeable about the accreditation criteria and A2LA policies and processes. There is a very rigorous process for qualifying assessor candidates to work for A2LA. A background check is performed, then new assessors are evaluated first as technical assessors on a team assessment, and then as the leader of an assessment team. The evaluation is based on ISO 10011-2 ("Guidelines for auditing quality systems - Part 2: Qualification criteria for quality systems auditors") (*5*). All new assessors and those needing refresher training are required to successfully complete a three day A2LA Assessor Orientation Course. Approved assessors are re-evaluated at least every three years. The assessors' written reports are also evaluated by the Accreditation Council, and laboratories are given the opportunity to evaluate the assessor who assessed their laboratory. Assessors are paid on a graduating scale up to $650 per eight hour day. All of their assessment expenses are reimbursed as well.

Advisory Committees. A2LA sets up advisory committees for certain fields of testing or program areas if advice is needed beyond that which can be obtained from existing consensus standards writing organizations or industry committees, such as ASTM. Each advisory committee provides advice on the development of program requirements and the interpretation and/or amplification of ISO/IEC Guide 25 requirements for a particular field(s) of testing. A set of bylaws governs the operation of each committee. The Chairman of each advisory committee becomes a member of the Criteria Council and each committee reports to the Criteria Council. Presently A2LA has advisory committees representing the Environmental, Construction Materials/Geotechnical, Transportation, Measurement and Reference Materials interests. Advisory committees meet as often as needed but at least annually.

Staff. A2LA maintains qualified paid personnel necessary for the effective operation of the Association. Management includes the President, who reports to the Board of Directors, Vice President, Information System Manager, Business Development Manager, Financial Manager and Technical Manager. Additional qualified staff are employed to support the laboratory accreditation operations of the Association. Staff presently numbers 22.

Domestic and International Activities

As the need for global acceptance of test data increases, A2LA's activities on the domestic and international fronts escalate as well.

International. A2LA presently has bi-lateral mutual recognition agreements with HOKLAS (Hong Kong), IANZ (New Zealand), NATA (Australia), SINGLAS

(Singapore) and SCC (Canada). However, A2LA is committed to developing multilateral agreements with counterparts in other countries. This permits United States laboratories accredited by A2LA to be recognized by international parties to the multi-lateral agreement as being competent within the accredited laboratories' Scopes of Accreditation. Such agreements reduce cost to United States manufacturers by allowing them to test their products once in a U. S laboratory with the results accepted in multiple foreign markets.

A2LA has completed the evaluation process and has been accepted into the Asian Pacific Laboratory Cooperation (APLAC) multi-lateral agreement. This evaluation included an on-site evaluation of A2LA's operations to the ISO Guide 58 criteria by a team of four accreditation peers and the witnessing of four ISO Guide 25 A2LA assessments. Seven accreditation bodies signed the initial APLAC multilateral agreement. These included A2LA (USA), CNLA (Chinese Taipei), HOKLAS (Hong Kong), IANZ (New Zealand), NATA (Australia), SINGLAS (Singapore) and NVLAP (USA). The signing took place in Japan in November 1997. There are an additional fourteen accreditation bodies who have signed memoranda of understanding toward eventually signing this multilateral agreement.

A2LA is also actively pursuing a multilateral agreement with the European Cooperation for Accreditation of Laboratories (EAL). There are seventeen member countries making up EAL including Belgium, Denmark, Finland, France, Germany, Ireland, Italy, Netherlands, Norway, Spain, Sweden, Switzerland, United Kingdom, Australia, Hong Kong, New Zealand and South Africa. A pre-evaluation has been completed and a full evaluation similar to the APLAC approval process will occur in the second quarter, 1998. Of particular concern to EAL is calibration measurement traceability back to a national standard for calibrations supporting testing laboratories. A2LA's policy requiring accredited testing laboratories to use accredited calibration laboratories where possible is a means to demonstrate that calibrations performed by US calibration laboratories are indeed traceable to a national standard of measurement.

Other regional agreement cooperations are being established. Argentina, Brazil, Chile, Columbia, Peru, Venezuela and the United States (A2LA and ANSI) are the member countries of the Inter-American Accreditation Cooperation (IAAC) which have recently signed a Memorandum of Understanding to cooperate toward the signing of a multi-lateral agreement. The North American Calibration Cooperation (NAAC) and Southern Africa Accreditation Cooperation (SAAC) are establishing themselves as well.

Domestic. One of the strongest supporters of A2LA has been General Motors. They recognize A2LA accreditation of their supplier laboratories, mostly in the mechanical and chemical testing areas, but now also in environmental testing. Similarly, companies like Shell Oil have come to rely on our accreditation in the environmental area in lieu of doing their own assessments. The automotive industry is also requiring A2LA accredited test data from commercial testing laboratories instead of QS9000 registration.

On a federal government level, A2LA is approved by NIST to accredit fastener testing laboratories per the Fastener Quality Act (PL 01-592) (*6*). Additionally, the Federal Communication Commission has recognized A2LA to accredit laboratories performing electromagnetic compatibility testing of computer peripherals. The US

Environmental Protection Agency recognizes A2LA for environmental Lead (Pb) testing. The Federal Highway Administration accepts test data from A2LA accredited laboratories performing construction materials and geotechnical testing of highway and airport projects.

Several State EPA offices also individually recognize A2LA accreditations of environmental testing laboratories. A2LA is also recognized informally by numerous municipalities and industry groups.

ISO/IEC Guide 25 (1990)

All laboratories accredited by A2LA are required to comply with ISO/IEC Guide 25 (1990), "General requirements for the competence of calibration and testing laboratories". In this Guide attention is paid to the activities of both calibration and testing laboratories and account is taken of other requirements for laboratory competence such as those laid down in the OECD Code of Good Laboratory Practice (GLP) (*7*) and the ISO 9000 series of quality assurance standards. Additional program requirements (specific criteria) for specific fields of testing (Environmental) or special programs (Fasteners) which are necessary to meet particular user needs (e.g., U.S. EPA, Fastener Quality Act PL 101-592) complement these general requirements in particular areas.

ISO Guide 25 is recognized on an international level as the appropriate standard for determining the competency of a laboratory to perform specific tests or types of tests, or calibrations. Guide 25 is a balanced standard addressing quality system requirements of ISO 9000 and the technical requirements needed to perform testing or calibration. The following criteria elements are included in ISO Guide 25:

- Organization and Management
- Calibration and Test Methods
- Quality System, Audit & Review
- Handling Calibration & Test Items
- Personnel
- Records
- Accommodation and Environment
- Certificates and Reports
- Equipment and Reference Materials
- Subcontracting Calibration or Testing
- Measurement Traceability and Calibration
- Outside Support and Services
- Complaints

A2LA publishes the ISO/IEC Guide 25 requirements with explanatory notes as Part A in the A2LA General Requirements for Accreditation (Green Booklet) (*8*). The Green Booklet also contains the Conditions for Accreditation (Part B) and the Accreditation Process (Part C).

A2LA Accreditation Process

Application. A laboratory applies for accreditation by obtaining the application package from A2LA headquarters and completing appropriate application sheets. All applicants must agree to the set of conditions for accreditation, pay the appropriate fees and provide detailed supporting information on:

- Scope of testing in terms of field(s) of testing, testing technologies, test methods, and relevant standards;
- Organization structure; and
- Proficiency testing.

Applicants must also provide their quality manual references which address the documentation requirements of ISO Guide 25 and provide a matrix of the technical training of their laboratory personnel.

On-site Assessment. Once the application information is completed and the appropriate fees are paid, A2LA headquarters staff identifies and tentatively assigns one or more assessors to conduct an on-site assessment. The laboratory has the right to ask for another assessor if it objects to the original assignment. Assessments may last from one to several days.

Assessors are given an assessor guide and checklists to follow in performing an assessment. These documents are intended to ensure that assessments are conducted as uniformly and completely as possible among the assessors and from laboratory to laboratory.

Before the assessment is conducted, the assessor team requests copies of the quality manual and related documentation (i.e., SOPs related to Guide 25 requirements) in order to prepare for the assessment. The quality manual and related documentation must be reviewed by the assessor team before the on-site assessment can begin. This review is done ideally before the assessment is scheduled. Upon review of submitted documentation, the assessor(s) may ask the laboratory to implement corrective action to fill any documentation gaps required by Guide 25 before scheduling the assessment. A pre-assessment visit may be requested by the laboratory as an option at this point to enhance the success of the full assessment. Prior to scheduling the full assessment, the assessor reviews the draft scope(s) to determine the tests to possibly witness and checks on the availability of the technical personnel who perform the tests. An assessment agenda is provided by the assessor.

The full assessment generally involves:

- An entry briefing with laboratory management;
- Audit of the quality system to verify that it is fully operational and that it conforms to all sections of ISO/IEC Guide 25, including documentation;
- Interviews with technical staff;
- Demonstration of selected tests or calibrations including, as applicable, tests or calibrations at representative field locations;

- Examination of equipment and calibration records;
- A written report of assessor findings; and
- An exit briefing including the specific written identification of any deficiencies.

The objective of an assessment is to establish whether or not a laboratory complies with the A2LA requirements for accreditation and can competently perform the types of tests or calibrations for which accreditation is sought. However, when accreditation is required to demonstrate compliance with additional criteria which may be imposed by other authorities, such as in the case of U.S. EPA, the A2LA assessment will include such additional criteria. A2LA has accredited a number of testing laboratories to the ISO Guide 25 requirements, but in addition has also assessed and accredited these laboratories to the OECD GLPs, at the request of the laboratories. Assessors may also provide advice, based on observations or in response to questions, in order to help the laboratory improve its performance.

Deficiencies. During the assessment, assessors may observe deficiencies. A deficiency is any nonconformity to accreditation requirements including:

- A laboratory's inability to perform a test or type of test for which it seeks accreditation;
- A laboratory's quality system does not conform to a clause or section of ISO/IEC Guide 25, is not adequately documented, or is not completely operational; or
- Laboratory does not conform to any additional requirements of A2LA or specific fields of testing or programs necessary to meet particular needs.

At the conclusion of an assessment, the assessor prepares a report of findings, identifying deficiencies which, in the assessor's judgment, the laboratory must resolve in order to be accredited. The assessor holds an exit briefing with top management of the laboratory, going over the findings and presenting the list of deficiencies (deficiency report). The authorized representative of the laboratory (or designee) is asked to sign the deficiency report to attest that the deficiency report has been reviewed with the assessor. The signature does not imply that the laboratory representative concurs that the individual item(s) constitute a deficiency. The laboratory is requested to respond within one month after the date of the exit briefing detailing either its corrective action or why it does not believe that a deficiency exists. The corrective action response must include a copy of any objective evidence (e.g., calibration certificates, lab procedures, paid invoices, packaging slips and training records) to indicate that the corrective actions have been implemented/completed.

If the laboratory fails to respond in writing within four months after the date of the exit briefing, it may be treated as a new applicant subject to new fees and reassessment should it wish to pursue accreditation after that time.

It is entirely possible that the laboratory will disagree with the findings that one or more items are deficiencies. In that case, the laboratory is requested to explain in its response why it disagrees with the assessor.

A laboratory that fails to respond to all its deficiencies within six months of being assessed shall be subject to being reassessed at its expense. Even if the laboratory responds within six months, A2LA staff has the option to ask for reassessment of a laboratory before an initial accreditation vote is taken based on the amount, extent and nature of the deficiencies. The Accreditation Council panel also has the option to require reassessment of a laboratory before an affirmative accreditation decision can be rendered.

Proficiency Testing. Proficiency testing is a process for checking actual laboratory testing performance, usually by means of interlaboratory test data comparisons. For many test methods, results from proficiency testing are very good indicators of testing competence. Proficiency testing programs may take many forms and standards for satisfactory performance can vary depending on the field. An accredited laboratory must participate in method-specific proficiency testing related to its field(s) of accreditation if such programs are available. Requirements for proficiency testing are prescribed by A2LA depending on the applicant laboratory's requested scope of accreditation. Unless otherwise specified in program requirements documents, a laboratory must participate in proficiency testing for one test method on each of their Scopes of Accreditation twice per year at a minimum. Greater participation is encouraged, however. When proficiency testing programs are not available for a specific method, the laboratory should demonstrate proficiency with internal performance-based data.

Accreditation Decisions. Before an accreditation decision ballot is sent to Accreditation Council members, staff shall review the deficiency response, including objective evidence of completed corrective action, for adequacy and completeness. If staff has any doubt about the adequacy or completeness of any part of the deficiency response, the response is submitted to the assessor(s). The laboratory may be asked to respond further to ensure a successful Accreditation Council vote.

Staff selects a "Panel of Three" from the Accreditation Council members for voting. The "Panel of Three" selection takes into account as much as possible each member's technical expertise with the laboratory testing or calibration to be evaluated. The laboratory is consulted about any potential conflicts of interest with the Accreditation Council membership prior to sending their package to the Accreditation Council. At least two affirmative ballots (with no unresolved negative ballots) of the three ballots distributed must be received before accreditation can be granted.

Staff shall notify the laboratory asking for further written response based on the specific justification for one or more negative votes received from the panel. If further written response still does not satisfy the negative voter(s), a reassessment may be proposed or required. If a reassessment is requested by more than one voter, the laboratory is asked to accept a reassessment. If the laboratory refuses the proposed reassessment, an Accreditation Council appeals panel is balloted.

If accreditation is granted, the A2LA staff prepares and forwards a certificate and scope of accreditation to the laboratory for each enrolled field of testing and special program. The laboratory should keep its scope of accreditation available to show clients or potential clients the testing technologies and test methods for which it is accredited.

A2LA staff also uses the scopes of accreditation to respond to inquiries and to prepare the A2LA Directory.

Annual Review. Accreditation is for two years. However, after the first year of accreditation, each laboratory must pay annual fees and assessor fees and undergo a one-day surveillance visit by an assessor. This surveillance visit is performed to confirm that the laboratory's quality system and technical capabilities remain in compliance with the accreditation requirements. For subsequent annual reviews occurring after the renewal of accreditation each laboratory must pay annual fees and submit updating information on its organization, facilities, key personnel and results of any proficiency testing. Objective evidence of completion of the internal audit and management review is also required. If the renewal laboratory does not promptly provide complete annual review documentation, or significant changes to the facility or organization have occurred, a one-day surveillance visit and payment of the associated assessor fees is required.

Reassessment and Renewal of Accreditation. A2LA conducts a full on-site reassessment of all accredited laboratories at least every two years. Reassessments also are conducted when evaluations and submissions from the laboratory or its clients indicate significant technical changes in the capability of the laboratory have occurred.

Each accredited laboratory is sent a renewal questionnaire, well in advance of the expiration date of its accreditation, to allow sufficient time to complete the renewal process. A successful on-site reassessment must be completed before accreditation is extended for another two years.

If deficiencies are noted during the renewal assessment, the laboratory is asked to write to A2LA within 30 days after the assessment stating the corrective action taken. All deficiencies must be resolved before accreditation is renewed for another two years.

Conclusion

Third party laboratory accreditation such as that offered by A2LA as a means to determine product quality continues to grow. A2LA welcomes all interested parties to join and support the Association in its global efforts toward eliminating multiple assessments, removing trade barriers and achieving worldwide acceptance of test data.

Literature Cited

1. ISO/IEC Guide 2: 1986, General Terms and their definitions concerning standardization and related activities.
2. ISO/IEC Guide 25: 1990, General requirements for the competence of calibration and testing laboratories.
3. ANSI/ASQC Q9002: 1994, Quality Systems-Model for Quality Assurance in Production, Installation, and Servicing.
4. ISO/IEC Guide 58: 1993, Calibration and testing laboratory accreditation systems-General requirements for operation and recognition.

5. ISO 10011-2: 1992, Guidelines for auditing quality systems; Part 2: Qualification criteria for auditors.
6. Fastener Quality Act, Public Law 101-592 (as amended by P.L. 104-113): May 14, 1998.
7. The Revised OECD Principles of GLP: 1997.
8. A2LA General Requirements for Accreditation of Laboratories: January, 1997.

Chapter 8

Laboratory Accreditation Under the National Environmental Laboratory Accreditation Program for Laboratories and Field Sites Conducting GLP Studies in the United States: An Update[1]

William W. John

E.I. du Pont de Nemours and Company, P.O. Box 80038, Wilmington, DE 19880-0038

The Environmental Laboratory Advisory Board (ELAB) is a federally chartered advisory board whose mission is to provide advice and recommendations to EPA and the National Environmental Laboratory Accreditation Conference. A subcommittee of the ELAB was formed in 1996 to examine the potential impact of laboratory accreditation on the community regulated under the EPA Good Laboratory Practice Standards (40 CFR Part 160). This paper describes the activities of the ELAB GLP Subcommittee and summarizes the final recommendations concerning accreditation of EPA regulated GLP laboratories to the ELAB.

The Environmental Laboratory Advisory Board (ELAB) is a federally chartered advisory board (composed from within EPA and the regulated community) whose mission is to provide advice and recommendations to EPA and the National Environmental Laboratory Accreditation Conference (NELAC) about programs involving laboratories who submit data for state and federal agencies to use in various decision making processes. Based on issues concerning the activities of NELAC and their impact on the Good Laboratory Practice community, in April of 1996 a Subcommittee of the ELAB was formed to: 1) Characterize EPA Good Laboratory Practice (GLP) laboratory evaluation needs; 2) Evaluate alternatives to accreditation; 3) Examine implementation options; 4) Determine benefits of GLP accreditation to EPA and others; and 5) Determine how action by EPA would impact the Organization for Economic Cooperation and Development (OECD) programs and commitments. The Subcommittee developed 5 primary options for consideration by the ELAB: 1) Augment the current program and increase funds for the EPA monitoring program; 2) Develop a third party accreditation system for GLP laboratories; 3) Increase the value of the current sponsor monitoring program; 4) Develop a process within the National Environmental Laboratory Accreditation

[1]**ELAB GLP Subcommittee Executive Summary and Final Report** Prepared for the July 28, 1997, Annual Meeting of ELAB

Program (NELAP, the program which governs NELAC) to accommodate EPA GLP standards; and 5) Develop a registration list for EPA's Federal Insecticide, Fungicide, and Rodenticide Act (FIFRA) and Toxic Substances Control Act (TSCA) testing laboratories. A number of significant needs for intraagency, interagency, and international harmonization, as well as the impact of potential EPA action on these needs, was developed. The cost of the current EPA GLP programs (including the quality assurance oversight effort) was found to be significant and undervalued by much of the regulatory oversight community (especially international regulators). The options developed by the Subcommittee will vary in cost and implementation complexity as well as their ability to address the various needs expressed in this report. It is clear that more resources should be made available to EPA's GLP monitoring and compliance program if these needs are to be met. However, the cost of adding these new resources must be balanced with the cost of the current program. Any new processes must be value-adding and cost-effective for the entire industry (regulators and regulated community) if these changes are to be successfully implemented. The international efforts to harmonize GLP programs and standards must also weigh heavily in any changes to the current program. The Subcommittee thinks their options provide for some interim relief to the current shortcomings and also offers suggestions for longer term improvements to the EPA GLP compliance monitoring program. The Subcommittee recommends that: 1) The GLP issue be disengaged from the current NELAP activity and timeline; 2) Interim relief be obtained by implementing the simpler aspects of key options outlined in this report; 3) The rule-making process be utilized to facilitate a long-term solution to the current problems; and 4) The entire regulated community be drawn into the review and comment aspect of possible solutions to ensure that each facet of the new program is cost effective and value adding, and redundancy is minimized.

Background

A series of 1991-92 Office of Inspector General (OIG) reports (*1-3*) concerning EPA oversight of GLP laboratories that submit data to be used in Agency decision making were very critical of the amount of auditing being done by EPA and the universe of facilities being audited. The report suggested that a third party accreditation program might be a more effective way to manage the oversight responsibilities of the Agency. At this same time an EPA and state sponsored effort was underway to create a National Environmental Laboratory Accreditation Program to set standards and normalize performance of environmental monitoring laboratories submitting data to the Agency as well as to many state and local decision making bodies (*4, 5*). A Federal Register Notice in December 1994 (*6*) indicated all organizations that submit data to EPA would be included in NELAP, including those regulated under the GLP standards of 40 CFR part 160 (FIFRA) and 40 CFR part 792 (TSCA). As the GLP community began to interact with those developing NELAP, many questions were raised by representatives of the GLP community at the first National Environmental Laboratory Accreditation Conference (NELAC) in February 1995. The Environmental Laboratory Advisory

Board (ELAB), formed during 1995, was charged with the responsibility to advise EPA and NELAC concerning problem areas relating to the implementation of the NELAP program. Several ELAB subcommittees (including a GLP Subcommittee) were organized to examine specific work which needed to be done in order for NELAP to become fully functional. The ELAB GLP Subcommittee was formed during the first quarter of 1996.

On April 23, 1996, twenty-six individuals from different parts of the GLP regulated community (private sector, EPA, USDA, FDA) met via telephone conference to discuss their ELAB GLP Subcommittee charter and begin a process of developing options for consideration by the ELAB (see Table I for the list of Subcommittee members).

Table I. Members of the ELAB GLP Subcommittee

David Alexander U.S. EPA	Maureen Barge FMC	Fran Dillon Stewart Pesticide Regis. Assoc.
David Dull (co-chair) U.S. EPA	Jimmy Flowers Dow Elanco	Debi Garvin Pacific Rim Consulting
Clive Halder Bayer Corporation	Louise Hess Lancaster Laboratories	Wynn John (co-chair) DuPont Ag Products
Robert Kiefer Chemical Specialties Mfrs. Assoc., Inc.	John D. Kobland American Cyanamid Ag Products Res. Div.	Francisca Liem U.S. EPA
Doris Mason Rhone-Poulenc Ag	Ray McAllister ACPA	John McCann McCann Associates
Chris Olinger U.S. EPA	Patricia O'Brien Pomerleau CIIT	Mick Qualls Qualls Ag Laboratory
Roxanne Robinson A2LA	Gary Roy Allied Signal, Toxicology	Patricia Royal Quality Systems Consultants, Inc.
Fred Siegelman U.S. EPA	Paul Swidersky Quality Associates, Inc.	Lee West RDA & NAICC
Tammy White USDA, Ag Exp. Sta.	Stan Woollen U.S. FDA	
Ad Hoc		
Ted Coopwood EPA	George Fong Florida Environmental Admin.	
John Henshaw Monsanto	Jeanne Mourrain EPA-AREAL	

It was decided that basic to any option considered was the need to maintain the current GLP standards, to meet the needs of the interagency and international community, and to be cost effective for members of the GLP regulated community. With this charge in mind the ELAB GLP Subcommittee divided into three sub-teams to address each facet of this charge. Team 1 was to look at options for the larger team to consider. They were also to examine the current EPA GLP compliance program and use this as a guide to bridge from present practice to potential options for the future. Team 2 was to look at the needs of intergovernmental agencies (EPA, FDA, etc.) and those of the international community. Team 3 was to develop information from a cost/benefit perspective which could be used to evaluate the cost effectiveness of the options selected and finally recommended to the ELAB.

On June 3, 1996, the ELAB GLP Subcommittee received a notice from the Environmental Monitoring Management Council (EMMC, a high level management team within EPA) of the EPA expanding the charter of the ELAB GLP Subcommittee to include looking at the GLP needs of all FIFRA and TSCA programs and to:

- Characterize the GLP laboratory evaluation needs of the Office of Prevention, Pesticides, and Toxic Substances (OPPTS) and the Office of Enforcement and Compliance Assurance (OECA).
- Evaluate feasible alternatives to NELAP accreditation.
- Examine program implementation options (e.g., NELAC, private sector, federal government).
- Determine the benefits of GLP accreditation to EPA and others.
- Determine how potential actions would impact Organization for Economic Cooperation and Development (OECD) programs and commitments.

The ELAB GLP Subcommittee integrated this expanded charter into the existing teams and began to address each of the areas of focus with the intent of preparing a final report and recommendations to the ELAB. Reports of the three teams of the Subcommittee are summarized in the following paragraphs.

Original Charter Team Summaries

Team 1. Development of Options and Examination of Current EPA GLP Compliance Program. A list of laboratory evaluation needs was developed based on input from OPPTS and OECA representatives to the team. The laboratory evaluation needs are summarized as follows. Additional resources are needed to enable EPA to inspect over 2,000 laboratories generating data for EPA submission. The majority of the 2,000 identified laboratories actually participated in fewer than five submitted studies during the fiscal years 1993-1995 (these statistics are the most current that are available). The current inspection program does not prioritize facilities by size or number of studies conducted. A means to identify the facilities generating data for EPA submission is needed. OECA currently relies on the

OPPTS database of facility names and addresses taken from the cover pages of study reports. These names and addresses could be several years old since the work is often done long before the study report is submitted to the Agency for review. Timely evaluation of studies for which regulatory decisions are pending is desirable. OPPTS team members also indicated that it would be preferable to prioritize inspections of facilities with large numbers of studies underway as well as those performing long-term and field studies. Persons performing the inspections should be capable of identifying technically meaningful issues and providing OPPTS with feedback regarding the importance of these issues. This list of needs also addressed issues raised by the1991-1992 OIG reports. From this list of needs, Team 1 identified a set of 35 options which would meet various aspects of the OIG Report and also address the concerns raised around laboratory accreditation. Discussions with the entire Subcommittee concluded that several of these original options overlapped with certain aspects of other options, and eventually the option set was reduced to five which were then evaluated and developed further. The following is a general description of the final five options.

Option 1. Augmentation of the Current Program and Increased Funding and Resources. The existing EPA OECA GLP compliance monitoring program would be continued but initially augmented in Phase I by redefining the universe of the facilities to be inspected with focus on facilities with study directors and primary/major data generators. Subsequently, the option could be expanded with increased funding in Phase II to increase the frequency of compliance monitoring so that sites could be visited in a more timely manner (2-3 years is the current international standard). Resources for the expansion in Phase II would come from one of three proposed sources: A) An increase in the EPA funds directed to the OECA; B) An increase in FIFRA registration fees could target funds for EPA to conduct GLP inspections; or C) Funds could be obtained from an EPA OECA directed "GLP Inspection" fee. The increased frequency of compliance monitoring would be expected to increase the public confidence and international acceptance of the US EPA GLP programs.

Option 2. Third Party Accreditation for Good Laboratory Practice Standards. The development of a private third party accreditation program would be sanctioned by EPA for the purposes of inspecting and accrediting laboratories to GLP standards. Enforcement responsibilities would remain with the EPA. The program would include registration of laboratories, on-site inspections of the test site facility, along with technical and quality programs. A certificate would be issued for successful completion of the GLP compliance inspection, which would address international concerns and broaden market acceptance of the laboratory and data. This option could function as either a mandatory or a voluntary program depending on the method of implementation.

As the Accrediting Authority, the U.S. EPA OECA would establish a program to recognize third party accrediting organizations or bodies to provide laboratory accreditation to a GLP standard. Interested stakeholders, including third party accrediting bodies, sponsors, contract laboratories and others would help develop recommendations for the Program Description including: A) OECA's responsibility as the Accrediting Authority; B) Criteria for approving third party accrediting bodies; and C) Criteria for qualifying and training assessors.

Interested third party accrediting bodies would develop their GLP accreditation program based on these conditions. These programs would be reviewed by OECA who would sanction acceptable programs. Continued approval would depend on OECA's monitoring and periodic reapproval of the accrediting program. Accepted accrediting bodies could publicize their approval and existing GLP accrediting program and begin to accept applications and complete the accreditation process as described.

Option 3. Increased Value of Sponsor Monitoring Programs. The existing EPA GLP Compliance Monitoring Program would be continued with the addition of recognition for sponsors' (registrant's) current and ongoing GLP inspection programs. Even though in the current program sponsors have full accountability for the quality and integrity of the data they submit to the EPA, the EPA has full responsibility for all aspects of compliance monitoring. In this option, EPA continues their inspection/audit program in generally the same manner, but by recognizing current value in existing sponsors' GLP inspections of contract facilities, the OECA targeting scheme from the list of 2000-plus facilities would be altered. Sponsors (registrants) would have a new responsibility to report to the Agency each time they visited and evaluated a contract facility, preferably in an established electronic format.

EPA would retain the option to inspect any test site, but would prioritize their schedule to focus on regular inspections of sponsors, testing facilities with study directors and facilities generating the majority of the GLP data. By establishing a data base of sponsors' GLP inspections, EPA would be able to track the number of sponsors' inspections at subcontracted test sites. Using this information, they would prioritize their need for inspections at remaining test sites that generate only a small amount of the data. By supplementing their inspection schedules with recognition of sponsor' schedules, the EPA would be much more effective in adequately scheduling inspection of testing facilities that generate the majority of the GLP data.

Option 4. NELAP Accreditation for Good Laboratory Practice Standards. In this option the current federally-controlled EPA GLP Program, utilizing the current GLP standards, would be placed under the umbrella of NELAP as a parallel program and would operate independently of the other NELAP programs. Federal EPA inspectors would conduct priority GLP compliance inspections and data audits as well as participate in the activities of NELAC, with additional inspection support being provided by EPA approved third-party assessors.

The allocation of responsibilities would be as follows: EPA would continue to manage and direct the activities of the EPA GLP program, to maintain records derived from GLP study and laboratory evaluations, and to address both interagency and international harmonization, regulatory, and compliance issues. NELAC would provide the administrative support for the accreditation program. Funding of the program would be largely derived from inspection fees levied by NELAC and/or third-party accrediting group(s) for accreditation inspections/ assessments. In summary, NELAC would be responsible for facility accreditation while EPA would retain oversight responsibilities for the GLP Program.

Option 5. FIFRA/TSCA GLP Test Facility Registration. Facilities that intend to perform FIFRA and TSCA GLP studies for submission to EPA would be required to register their facility with EPA. Facility registration would involve an initial submission of information and documents from the facility for review to establish the basic profile for the facility. Documentation could possibly include: Description of size, organization, and capabilities of the facility; the organization, functions, and procedures of the quality assurance unit; general description of instruments and equipment used at the site; and the number and areas of expertise of staff. It might also include a current list of standard operating procedures; resumes, curriculum vitae, and training records of key personnel; floor plans of the facility; and a current master schedule. On a periodic schedule, facilities would be required to resubmit certain documents and information.

The Agency or a designated third party contractor(s) would audit the submitted documents. Registration would not confer approval. Facilities with corrected minor GLP deficiencies would be provisionally registered, while facilities with major GLP deficiencies would be targets for inspection. Periodic submission of the facility's master schedule would be required and would provide a means of monitoring work intended for submission to the Agency. This would allow OECA to prioritize its inspections and be able to conduct in-life audit reviews of on going studies. To remain on the registration list, a submitter would need to continue to remain in GLP compliance verified by an EPA facility inspection audit.

Team 2. Interagency and International Issues Concerning Laboratory Accreditation. The following issues were identified:

U.S. Interagency Issues Pertaining to U.S. EPA Lab Accreditation - FDA Position Statement. Departments, Agencies and Administrations outside EPA potentially affected by a GLP accreditation program include the USDA (IR-4 Program) and the FDA GLP. While internally, USDA does not have GLP requirements, USDA programs, such as IR-4, that collect and submit data to EPA in support of pesticide registration do require GLP compliance programs as part of their funding requirements. The FDA, on the other hand, has a well-established GLP program. The outcome of the debate on developing a national GLP accreditation program has the greatest impact on this program.

Since 1978, the FDA has had a program for inspecting GLP laboratories conducting non-clinical safety studies for pharmaceuticals, veterinary products, and medical devices (frequency every 2-3 years). Such studies are conducted and reported in accordance with the GLP regulations found in 21 CFR part 58. There are currently no plans by the FDA to adopt an accreditation approach to regulate GLP laboratories. The program of inspections and data audits, currently in place at the FDA, provides the necessary level of data quality and integrity with a minimal outlay of resources.

In developing its approach for regulating these laboratories, the FDA considered several options, including a third party accreditation program. The FDA concluded that a program of regular laboratory inspections and data audits, conducted by FDA personnel, was the most cost effective and efficient means to ensure the quality and integrity of data submitted to the FDA. The FDA reached this conclusion, in part, based upon its decision to include in the regulations a requirement that each laboratory appoint an independent quality assurance unit, as an internal monitoring process. This self-regulation approach was favored by the FDA as the least burdensome to industry and most efficient for FDA oversight. The advantages of the FDA approach to regulate non-clinical safety testing laboratories is recognized domestically and by other agencies of the U.S. government and internationally, including the EPA and OECD.

Implementation of an accreditation program by a third party would entail the added expenditure of resources to establish an infrastructure of training, oversight and additional regulations. There has been no information presented to the FDA at this point to suggest any justification for this added expense, nor does the FDA have any indication that its current program has been ineffective (*7*).

International Issues Pertaining to U.S. EPA and the OECD GLP Programs. The development of a United States GLP standard by the FDA in the late 1970s prompted interest in GLP on the part of other OECD Member countries in order to ensure continued acceptance of their data in the large U.S. market. OECD's involvement flowed logically from a principle purpose of all of its programs--- the avoidance of nontariff trade barriers between OECD Member countries as a consequence of national regulatory programs. It is frequently stated that the goal of the OECD program is the "international harmonization" of GLP requirements. In general, the OECD Member Countries with national GLP programs have adopted the OECD Principles of GLP as the basic standard, as required by the 1981 Council Act. This is especially true for the 15 member states of the European Union, (whose standard is the OECD Principles verbatim), Japan (MHW, MAFF, MITI), the United States (FDA and EPA), and Switzerland. In general, there is a very high degree of harmonization among these countries. Newer programs based on GLP are being developed in Canada, Mexico and Brazil.

Equally relevant to analyzing the impact and conditions of a U.S. GLP accreditation program is the evaluation of existing bilateral agreements and Memoranda of Understanding (MOU) between the U.S. and OECD Member countries. These agreements reiterate provisions for meeting the Mutual Acceptance of Data Decision and goals, including promotion of data acceptance and reciprocity among participating countries, and continued cooperative relationship between countries. Requirements can be summarized into four general conditions: 1) Adherence to standards of GLP based on national GLP programs and the OECD Council Recommendations and Decisions; 2) Mutually consistent national programs, including periodic (approximately every two years) inspections by trained government inspectors, (or government sanctioned programs); 3) National compliance procedures, including the notification of laboratories with observed deficiencies and requirements for corrective action; and 4) Periodically, providing the signatories with names and addresses of non-clinical health and environmental safety laboratories operating within the country, the dates of government or government sanctioned inspections, and current GLP compliance status.

None of these requirements either negate or promote the concept of developing a U.S. GLP Laboratory accreditation program. Critical, however, to evaluating the impact of accreditation on the U.S. EPA GLP program is the preamble to the document entitled "Revised Guide for Compliance Monitoring Procedures for Good Laboratory Practices." The preamble states that "Member countries will adopt GLP Principles and establish compliance monitoring procedures according to national legal and administrative practices..." Thus, it would appear evident that EPA could establish a third party accreditation program as long as EPA played an appropriate role in establishing and overseeing the program. This conclusion is consistent with programs already in place in several European countries.

Team 3. Cost/Benefit Analysis of Current Programs to Industry and Proposed Options. A survey was developed and distributed to the EPA GLP community in an attempt to better understand the cost of the current GLP regulations to the regulated community. This survey was conducted in an effort to determine the impact of GLP on the cost of research and to break-out the cost of the Quality Assurance Unit (QAU) as it monitors these programs. Approximately 900 Cost/Benefit Survey forms were mailed to members of the Society of Quality Assurance (SQA) (400, members operating under EPA GLP regulations), National Association of Independent Crop Consultants (NAICC) (120), Chemical Specialty Manufactures Association (CSMA) (300), and American Crop Protection Association (ACPA) (~80). These organizations were to pool their results into a single response for the entire organization. Fifty two responses were returned (Sponsors, 16; Contract Labs, 14; Field Cooperators, 16; and Others, 6). The small response was insufficient to provide a reliable estimate of the total cost of GLP regulations to the industry. However, the cost of the QAU did provide some insight into the cost of the Quality Assurance portion of the GLP. The average cost of a QA professional from the companies represented in the survey was approximately $70,000 per year (this cost would include benefits, travel, and QA program cost in addition to salary). Since there was a higher response of sponsors relative to independent QA respondents, this number may be an over estimate of the industry average; however, it should not be drastically wrong. If this number is multiplied by the approximate 400 members of SQA associated with EPA programs, then it is clear that the current direct cost to industry for GLP QA programs approaches $30 million annually. Additional indirect costs (i.e., archiving, training, etc.) drive this value even higher. This number becomes particularly significant when it is realized that the OIG Reports issued between 1991 and 1992 did not give any consideration/credit for the impact that EPA regulated industry QA programs have on data quality. The FIFRA and TSCA testing industry GLP QA program effort must be considered in whatever final decision is reached in the current oversight/monitoring debate if an acceptable cost-effective revision is to be successfully implemented.

How the Options Address the Expanded Charter

The charter of the Subcommittee was expanded by the EMMC at such a time (June 1996) that consideration was given to the new charter throughout the work of the Subcommittee. More detailed responses to the individual options relative to the expanded charter objectives were provided to the ELAB. An overview of the findings relative to the individual aspects of the expanded charter are presented below:

Characterize the Laboratory Evaluation Needs of OPPTS and OECA Programs. The various needs identified were characterized as noted below.

- Need to know who is currently doing the work (universe of laboratories). There are over 2000 laboratories currently supplying data for EPA review.

- Need to know when the work is being done. In-life audits are far more valuable and much less controversial than postmortem findings. This is true both for determining the quality of the work being done and improving the quality for future work.

- Need to know the level of compliance of the study during the data review phase, not after the tolerance has been set and a registration granted.

- Need technically trained inspectors who can identify meaningful technical issues and assess their impact on the study (administrative problems are less of an issue than technical problems) and the review process.

- Need to have critical phase and timely ongoing audits for long-term studies.

- Absolute requirement for necessary resources to conduct timely audits and to provide reasonable monitoring oversight.

- Need to be able to balance work so heaviest data suppliers get reasonable oversight, but that all facilities are audited in a timely, regular manner. This is particularly critical as international harmonization activities are increasingly changing the compliance arena.

Evaluate Feasible Alternative to Accreditation. Three nonaccreditation options were identified which would meet various aspects of the needs. They are:

- Augmentation of current programs and increased funding.
- Increased value for sponsor monitoring programs.
- Laboratory registration.

Examine Program Implementation Options. Specific examples of how each of the options discussed would be implemented were presented to the ELAB. There are general issues which should be resolved no matter which option is ultimately selected. They are:

- Define the standard which will be used for the future monitoring program. This may be as basic as ISO vs. GLP standards, revised GLP standards to meet new OECD GLP Principles, or develop a new standard for an accreditation system (if necessary). The Subcommittee sees the greatest value in amending the current US GLP to meet the harmonization standard of the revised OECD GLP.

- Reaffirm the federal program basis for FIFRA and TSCA programs. This seems to go without saying, but continues to be a key part of the debate relative to NELAP.

- No matter which option or program is pursued in the future there may be a requirement for legislative changes to FIFRA and TSCA as well as amendments of the current EPA GLP standards to facilitate implementation of the new program.

- There will be a need to provide training and certification of new auditors who will be required to meet the expanded monitoring requirements.

- The resources for the overall program should be reevaluated. If the current resources (dollars, manpower, and time) are not adequately meeting the needs, then they should be examined as part of the whole process. If new costs are to be added, every effort must be made to make certain there is not a redundancy in what is being done by the EPA and what is required by the industry. Each step must be value adding.

Determine the Benefits of Accreditation to EPA and Others. Potential benefits of accreditation are listed here; detailed benefits **and** disadvantages were presented to the ELAB:

- Provides increased frequency of inspections, while allowing OECA to retain its authority and enforcement responsibilities.
- Facilitates OECA's focus on data audits.
- Provides an "approved" universe of laboratories.
- Facilitates integration of regulatory and commerce issues, and streamlines administrative duties.
- Meets international (OECD GLP) requirements.
- Provides greater international acceptance of laboratory testing programs and data.
- Financially self-sustaining, fee would be assessed to cover program cost.

Determine How Potential Actions Would Impact OECD Programs and Commitments. The OECD GLP principles were revised and were issued early in 1998. These standards are geared to drive international harmonization of regulatory work and requirements. The US EPA GLPS will need to be amended if we are to meet the new international harmonization standards. The new standard will help determine the potential value of each of the options developed at this time.

Discussion

In summary, the Subcommittee has developed five primary options for consideration by ELAB. These options will vary in cost and implementation complexity as well as their ability to address various interagency and international needs expressed earlier in this document. Phase I of option 1, option 3, and option 5 will allow the Agency to augment the existing compliance monitoring program with minimal resource drain and added cost. But, the expectation is that these options, as stand alone options, are unlikely to meet all of the concerns of the international community or the OIG comments concerning frequency of EPA GLP compliance monitoring. Additional resources (both manpower and capital) will be required in order to satisfy these more complex concerns. The decision tree depicted in Figure 1 identifies the Subcommittee's preference for consideration of alternative, more comprehensive options, taking into account numerous relevant factors (complexity, cost, timing, value-adding potential, and ease of implementation) that are expanded upon throughout this chapter.

The ELAB should be advised that concurrent with this investigation of options to improve the current EPA GLP Compliance Monitoring program are the efforts to harmonize the GLP Standards internationally through revision of OECD

Figure 1. Decision Tree View of an Implementation Scheme for the 5 Options Developed by the ELAB GLP Subcommittee.

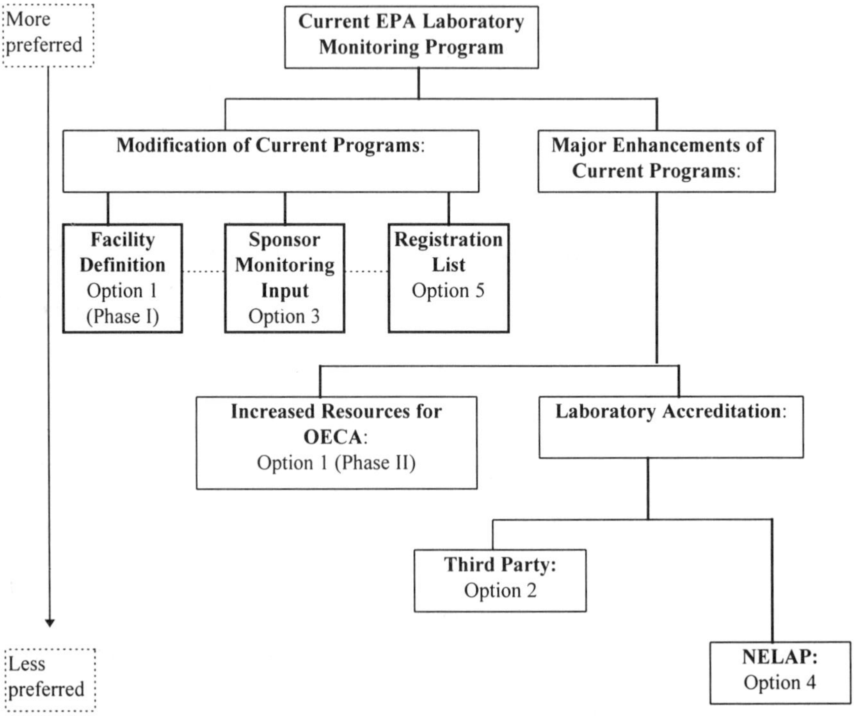

Principles of GLP followed by their adoption by OECD Member Countries. This harmonization initiative is likely to impact to some degree the options identified in this report (particularly the larger, more comprehensive phase II of option 1, option 2, and option 4). Since the revision efforts are fairly close to being realized, it is recommended that decisions covering phase II of option 1, option 2, and option 4 be deferred until after the new OECD GLP Principles are published and the harmonization activities are concluded.

Subcommittee Evaluation of Options. Since each of the options listed above addressed different issues or concerns raised by members of the Subcommittee, an effort was made to consolidate them into packages. Each option represented different approaches to meeting the needs (depending on the amount of time and resources to be applied to the problem). The options also spanned a number of potential activities which were very divergent in nature. Based on these considerations the Subcommittee concluded that the development of a decision tree view of the options (Figure 1) was more representative of their conclusions than reducing/condensing the option set further.

The Subcommittee considered complexity, cost, timing, value-adding potential, and ease of implementation and then through a multi voting process concluded that options 1 (phase I), 3, and 5 would be the easiest (fastest and least costly) to implement. The Subcommittee recognized that these three options by themselves or in combination were unlikely to address key international concerns such as frequency of auditing and/or certification. Option 1 (phase II), notwithstanding the difficulty in obtaining additional funding, has the greatest potential of addressing the most needs with the least disruption and cost. Option 2 may require more resources to implement. There was one overriding consensus within the Subcommittee and that was that option 4 was the least attractive of all the options because it posed the largest number of issues and constraints (detailed list provided to the ELAB). It is critical that any additional program cost be offset by value-adding benefits to industry for any of these options to be implemented successfully.

GLP Subcommittee Recommendations to the ELAB

1. Disengage the GLP issue from the NELAP activity and timeline. There are too many potential problems with this option relative to other options notwithstanding the differences in program needs, resources, etc.,
2. Focus immediately on implementing options 1 (phase I), 3, and/or 5 to augment the current OECA compliance monitoring program. Should this modification in concert with harmonization efforts with the OECD GLP Principles still not address the perceived deficiencies of the OECA compliance monitoring program, thereafter, consider on a longer-term scale, the value to be added by implementing other options identified in this report,

3. Utilize the rule making process to amend the US FIFRA/TSCA GLP standards to meet the new OECD Principles of GLP and for alignment with the many international harmonization efforts underway at the current time, and
4. Utilize the rule making process to include the entire regulated community in the review and comment discussion of possible solutions to ensure that each facet of the new program is cost-effective and value adding, and that redundancy is minimized.

ELAB Response to the Subcommittee Recommendations

The ELAB asked the Subcommittee to prepare a final list of options which reduces the 5 options presented in July to a simplified list consisting the following options:

1. No Change (the Subcommittee takes this to mean use the current processes in place at OPPTS/OECA to address the current issues);
2. Augmentation of the current program; and
3. Accreditation (the Subcommittee does not currently agree with the ELAB that third party and NELAP accreditation are the same option).

The Subcommittee is in the process of making this new proposal and will provide their final report to the ELAB at the interim meeting in January 1998.

Literature Cited

1. EPA-OIG E1PG1-11-00210140046, EPA's procedures for ensuring quality data under the GLP Program, 9/30/91.
2. EPA-OIG-E1PG1-11-0028-2400032, Alternatives for ensuring accurate laboratory data, 3/31/92.
3. EPA-OIG-E1EPF2-11-2100661, Cooperation and coordination problems limit effectiveness of GLP Program, 9/30/92.
4. EPA Final Report; CNAEL Committee, EPA 1992.
5. NELAC Constitution and By-laws; 1995.
6. 50 FR 66178, 12/2/94 - NELAC Program.
7. Personal Communication, Stan Woollen, US FDA, 1997.

Chapter 9

Good Laboratory Practice Regulations of the U.S. Environmental Protection Agency

Our Mission: Past, Present, and Future

Francisca Liem and Mark J. Lehr

Laboratory Data Integrity Branch, Office of Compliance, Environmental Protection Agency, Washington, DC 20460

To one degree or another scientists have always worked within a set of principles which they consider to be good laboratory practices. American chemical testing laboratories took particular notice when the United States Environmental Protection Agency (EPA) revised its national testing standards known as the Good Laboratory Practice (GLP) regulations (40 Code of Federal Regulations (CFR) sections 160 and 792) in 1989. These new regulations, intended to govern testing for pesticides registration and toxic chemical manufacturing, motivated a large segment of the scientific community to standardize the concept of "good laboratory practice."

The GLP regulations were initially enacted in 1979 to cover health effects of drug testing and food additives for the U.S. Food and Drug Administration. In 1984 the EPA adopted the regulations to include the effects of pesticides on human health and domestic animals for studies to support marketing permits for pesticides. The EPA GLPs also covered all testing on toxic substances. The GLP regulations were revised in the fall of 1989 (*1*), to include all other chemical testing required by EPA for pesticides and toxic substances, most notably, environmental and chemical fate and ecological effects testing. Several principles were either introduced or formalized by the GLP standards regulations which had not always been utilized by laboratories working under what they considered to be good laboratory practices. These new aspects include the concepts of a formal study director, a defined quality assurance unit (QAU), a written and approved study protocol, standard operating procedures (SOPs), and final report requirements.

Since formalizing these standards, interest in the GLP regulations has grown throughout the United States and is now expanding beyond its boarders to countries around the world. Nations in Europe, Asia and the Pacific, Latin America, and North America are actively implementing quality assurance programs based on the GLP regulations. The Organization of Economic Cooperation and Development (OECD), an intergovernmental organization consisting of twenty-four industrialized nations, recognized the need for a program aimed at ensuring valid, high quality test data. The OECD determined that globally accepted data must be based on a program comprised of GLP principles, including a mechanism for monitoring and assuring compliance with the adopted GLP principles. This harmonization and broad acceptance of the basic concepts of GLP was an important step toward mutual acceptance of data. An international group of experts on GLP standards was established and work began developing a set of guidelines to govern the conduct of laboratory studies world wide. This effort resulted in the development of the first OECD GLP principles which were established in 1981 (*2*). Today OECD continues to refine its quality assurance guidelines using GLP Standards as its cornerstone. OECD has prepared revised guidelines which were published in January 1998 (*3*).

In the United States, EPA's GLP program is directed by the Laboratory Data Integrity Branch (LDIB). The LDIB is located within the Office of Enforcement and Compliance Assurance under the Agriculture and Ecosystems Division. LDIB's primary mission is to assure the quality and integrity of studies submitted to the agency in support of application for research and marketing permits for pesticide products. EPA accomplishes this mission by conducting laboratory inspections and data audits to assure compliance with the GLP regulations. The past thirteen years have shown remarkable improvements in the quality of the data submitted to the EPA as a result of the GLP regulations. Laboratory enhancements in the areas of quality assurance/quality control, record keeping, and accountability have lead to a heightened level of scientific quality which is recognized throughout the world. The GLP regulations are the driving force behind this movement and the EPA is committed to continuing this trend.

The EPA has continued to improve its GLP inspection program by giving inspectors a greater role in determining the fate of inspection reports which has greatly streamlined our ability to refer and close cases. Additionally, EPA has managed to increase its efficiency in the field, thus allowing its inspectors to conduct more inspections with a smaller inspection staff. In fiscal year FY 97 the EPA performed 127 inspections which included 480 data audits of studies already submitted to the EPA. The EPA currently estimates that approximately 1,400 laboratories are performing studies in accordance with the GLP regulations. Of these facilities, nearly 550 are analytical chemistry laboratories, 309 are performing field testing, 264 are toxicology facilities, 69 are insecticide efficacy, and 19 are antimicrobial laboratories. To accurately characterize and target these facilities the EPA has made several improvements to its inspection targeting data base known as Laboratory Inspection and Study Audit, or LISA. Recent modifications made to LISA have greatly enhanced its capabilities. Last year EPA staff members performed inspections at 84 testing facilities which had never before been audited. These "new testing facilities" accounted for 66% of the total inspections performed last year. Analytical chemistry laboratories were the most frequently inspected facilities in 1997 making up 52.0% of the total number of inspections. Field testing sites were second at 17.3%, and toxicology laboratories third at 16.5%. In 1997 antimicrobial, environmental effects and insecticide efficacy laboratories were the least inspected laboratories at 7.1%, 3.9%, and 3.1% respectively. In addition to using LISA for scheduling inspections, EPA's LDIB also targets facilities based on requests made by the Office of Pesticide Programs (OPP). OPP utilizes LDIB inspectors to address questions that arise during the pesticide review process.

By carefully monitoring compliance trends from year to year, EPA is able to focus its time and resources in areas of the regulated community with the poorest compliance rate. In FY 97, EPA found that analytical testing facilities, more specifically product chemistry labs, were the least GLP compliant sector. EPA inspectors found that only 18 out of 66 analytical facilities, or approximately 27%, were fully compliant with the GLP regulations. Sponsor-run laboratories performing product chemistry testing were the most violative of these facilities, and consequently, the most frequented, accounting for 44 of the 66 inspected chemistry laboratories. By contrast, percent compliance rates range from 55% for field testing facilities to 33% for antimicrobial laboratories. Reasons for noncompliance in product chemistry laboratories will vary greatly from one facility to the next; however, there are possible explanations for poor regulatory performance in this sector. Product chemistry laboratories are typically found in pesticide formulation/manufacturing plants which focus their efforts in areas of production and packaging. While issues concerning quality assurance/quality control are very important to these facilities, attention to regulatory requirements sometimes finds itself taking a backseat. Additionally, the GLP regulations allow for certain exemptions under physical and chemical characterization studies (40 CFR §160.135). Because of these exemptions, product chemistry facilities sometimes perceive the GLP regulatory requirements as not necessary or unimportant. In many cases, product chemistry work is contracted "out-of-house" to facilities not equipped to handle GLP regulatory requirements. It should be restated, however, that analytical laboratories account for the largest segment of testing facilities performing GLP work in the United States.

After an inspection has been completed, it is the responsibility of the inspector to write a legally defensible report in support of Agency enforcement efforts. Violative cases are referred to either the Office of Regulatory Enforcement (ORE) for regulatory concerns or to the Office of Pesticide Programs (OPP) for scientific issues. In some instances reports are referred to both ORE and OPP if regulatory and scientific concerns are raised by the inspection team. Table I shows a comparison, by discipline, of old vs. new facilities inspected during FY 97. In addition, the table compares the number of inspections to the number of facilities referred, and the compliance rate of old vs. new facilities. Also provided are the total percent compliance rates for each discipline.

Table I

Compliance Rate FY 1997 Referrals, Old vs. New, and Disciplines								
Discipline	Number Inspected		Number Referred		Number in Compliance			
TESTING LABS	*OLD*	*NEW*	*OLD*	*NEW*	*OLD*	*NEW*	*TOTAL*	
Toxicology	20	1	2 (10%)	1 (100%)	18 (90%)	0 (100%)	18 (86%)	
Analytical Chemistry	12	54	1 (18%)	26 (48%)	11 (92%)	28 (52%)	39 (59%)	
Antimicrobial Efficacy	3	6	1 (33%)	2 (33%)	2 (67%)	4 (67%)	6 (67%)	
Field Sites	6	16	1 (17%)	0 (0%)	5 (83%)	16(100%)	21 (96%)	
Environmental Effects	2	3	1 (50%)	0 (0%)	1 (50%)	3 (100%)	4 (80%)	
Insecticide Efficacy	0	4	0 (0%)	4 (100%)	0 (0%)	0 (0%)	0 (0%)	

Total number of inspections: 127

FY 1997 marked the second year the EPA has utilized its Inspection Observation Form, also known as the 038. The 038 provides the inspected facility with instant written observed regulatory findings made by the inspection team during the audit. The 038 also provides EPA management with real-time information which can be used to better address problem areas within the regulated community as they occur. Table II lists the inspectors' findings observed during FY 97 in each subpart and their occurrence by percent.

Table II

Inspector Findings Inspection Observation Forms (038) FY 1997		
GLP Subparts	Number of Findings	Percent Findings
A - General Provisions	9	7%
B - Organization and Personnel	54	43%
C - Facilities	10	8%
D - Equipment	17	13%
E - Testing Facility Operation	21	17%
F - Test, Control, and Reference Substance	23	18%
G - Protocols for the Conduct of a Study	46	36%
J - Records and Reports	4	3%

As the table shows, Subpart B - Organization and Personnel, and Subpart G - Protocol for and Conduct of a Study were the most violative areas of the GLP regulations during 1997. Listed below are the most common observations made by EPA inspectors from these two subparts.

Subpart B - Organization and Personnel

- Lack of Independent Quality Assurance Unit (QAU)
- Lack of Quality Assurance (QA) inspections
- Problems associated in the routing of QA reports
- Lack of QA records
- Lack of/incomplete QA records
- Lack of/incomplete master schedule
- Final report does not match the raw data
- Lack of a study director
- Unforeseen circumstances not documented/reported
- Lack of technical training and records

Subpart G - Protocol for and Conduct of a Study

- Lack of raw data
- Data missing
- Lack of a signed and/or approved protocol
- Protocol missing required GLP elements
- Lack of signed protocol changes
- Incorrect calculations
- Records in pencil, not initialed and/or dated

As previously mentioned, organizational and technological changes have enabled inspectors to work smarter and faster resulting in stronger cases. These positive changes manifested themselves on April 24, 1997, when six Federal Insecticide Fungicide Rodenticide Act (FIFRA) GLP enforcement initiatives were announced by the Office of Regulatory Enforcement (ORE) resulting in nearly $70,000 in fines. Among the facilities fined were three product chemistry laboratories, two toxicology laboratories, and two efficacy facilities. The resulting GLP violations included in the enforcement action are listed as follows:

Subpart B - Organization and Personnel

- QAU failure to assure the final report accurately reflected raw data
- QAU failure to prepare and sign a statement in the final report, specifying inspection dates and findings reported to management and the study director
- QAU failure to submit written status reports to management
- QAU failure to maintain a copy of the master schedule
- QAU failure to maintain written records describing responsibilities/ procedures applicable to QAU
- Study director failure to assure all experimental data were accurately recorded and verified

Subpart J - Records and Reports

- Failure to describe in final report all circumstances that may have affected the quality/integrity of the data
- Failure to include in the final report all required GLP elements
- Failure to retain all raw data, documentation, records, protocols, specimens and final report of the study
- Failure to archive master schedule, receipt of test substance, test substance accountability form with the study files
- Failure to archive written approved test substance dose preparation instruction with study files

Subpart G - Protocol for and Conduct of a Study

- Failure to document changes to the protocol
- Failure to have an approved written protocol for the study
- Failure to record data directly, promptly, and legibly in ink
- Failure to initial and date data entries

Subpart E - Testing Facilities Operation

- Failure to follow Standard Operating Procedures (SOP) for test system care

FY 97 also brought changes to the types of inspections EPA routinely performs. In addition to conducting GLP regulatory inspections, the EPA began conducting audits of laboratories claiming non-GLP compliance for submitted studies supporting permits for pesticide products. EPA is concerned about the quality, integrity, and reproducibility of all studies submitted to the Agency for registration including studies claiming non-compliance. These inspections are performed under the purview of FIFRA Books and Records [40 CFR 169.2(k)] which requires all records supporting a registration to be retained as long as the registration is valid and the producer remains in business.

Among the top priorities of the EPA is to ensure compliance with environmental laws and regulations and to help ensure that good science is used in EPA decision making. Historically, EPA inspectors have played a key role in this effort by conducting rigorous inspections and identifying violations for subsequent enforcement actions. This approach has served the EPA well in the past in helping laboratories achieve compliance and providing the necessary deterrence to the regulated community. However, as the number of regulatory requirements and number of regulated entities has increased, it has become clear we need a more effective means to maintain this strategy. Promoting new innovative approaches in EPA's GLP program is essential to maximize compliance and encourage data quality and integrity within the regulated community. EPA is also working with universities around the nation to establish a quality assurance standards based curriculum designed to stress data integrity and the fundamentals of GLPs. EPA recognizes that maintaining the upward trend in data quality rests in the hands of future scientists. During the next year, EPA's targeting staff will continue to extend the range of its laboratory inspection data base, LISA. The EPA is dedicated to expanding our inspection universe by locating unidentified laboratories and reaching out to new regulated sectors. In addition, EPA has recently taken steps to make training manuals, SOPs, and GLP regulatory advisories more accessible by placing them on the Internet.

In supporting our primary mission of protecting human health and the environment, EPA will continue to stress the importance of good science. While enforcement and technical assistance will continue to remain the primary tool EPA uses to achieve and measure compliance, we are seeking other tools to promote and ensure data quality and integrity. The Agency must continue to identify environmental and health risks, analyze the underlining causes of noncompliance, and apply appropriate solutions. Developing and using new procedures to carry out this approach will place increased demands on field personnel. Today EPA inspectors must have sound technical skills and be capable of accurately conveying regulatory requirements. EPA will continue to support programs that promote compliance assistance, as well as further communication and outreach. It has always been our goal at the EPA to push the limits of quality and innovation, and this philosophy will continue into FY 1998. The EPA is dedicated to working as a partner with industry to adequately carry out this new approach.

Literature Cited

1. "Good Laboratory Practice Standards Under the Federal Insecticide, Fungicide, and Rodenticide Act Final Rule", *Code of Federal Regulations* Title 40, Part 160; *Federal Register* 54:58 (August 17, 1989) pp 34067-34074.
2. The OECD Chemicals Programme, OECD Publications, No. 72009, 1989.
3. OECD Series on Principles of Good Laboratory Practice and Compliance Monitoring, Number 1, OECD Principles on Good Laboratory Practice, ENV/MC/CHEM(98)17, 1998.

Note: Opinions presented in this paper reflect that of the authors and should in no way be perceived as official EPA interpretation.

Chapter 10

Proposed Changes in FIFRA GLPS: Impact on the Agricultural Industry

William J. Litchfield

Agricultural Products, E.I. du Pont de Nemours and Company, Wilmington, DE 19880–0402

Significant changes are being proposed (*1*) by the U.S. Environmental Protection Agency to the Good Laboratory Practice Standards (40 CFR Part 160). At least nine of these changes were also proposed by the American Crop Protection Association's GLP Work Group, which presented a list of its recommendations (*2*) to EPA in April 1996. A review will be given of the proposed changes, their impact on the U.S. agricultural industry, and the prospects for additional modifications in the future.

Driven by a mandate to reduce paperwork, the U.S. Environmental Protection Agency (EPA) is now proposing changes to its Good Laboratory Practice Standards (GLP or GLPS) Regulations that will consolidate 40 CFR Part 160 (FIFRA) (*3*) and 40 CFR Part 792 (TSCA) (*4*). As part of this effort, EPA is also proposing amendments that are intended to "streamline and ease compliance" while maintaining data integrity. EPA's Office of Enforcement and Compliance Assurance (OECA) has indicated that such changes are coming for the past few years, and the Office has been open to suggestions offered by various organizations. To solicit public comment, a draft document containing these changes will be published in the Federal Register during 1998.

The American Crop Protection Association (ACPA) has been active in this area since early 1994, when it organized a GLP Work Group and solicited comments on 40 CFR Part 160 from its eighty member companies. The GLP Work Group met frequently over the next two years to consider changes that could (1) improve the efficiency and cost effectiveness of FIFRA GLPS, (2) enhance international harmonization, and (3) adapt new technologies, while maintaining data integrity and public safety. Over fifty changes proposed by the GLP Work Group were presented to OECA in April 1996, and a few of these are similar to ones now being proposed by EPA. The GLP Work Group endorses the consolidation of TSCA and FIFRA GLPS and encourages public comment on the following EPA proposed amendments.

Proposed Changes and Impact

In Subpart A under Definitions, EPA has proposed essentially four changes. Considering the first two of these as they appear in a draft to be published in the Federal Register, (1) the definitions of carrier and test system include "air", and (2) the definition for the Quality Assurance Unit excludes "individual(s) directly involved with the conduct of the study." Both of these changes are relatively innocuous and should have little impact on industry-wide programs to comply with GLPS. A third proposed change, however, affects the definition of raw data, so that it includes "any original data captured electronically or by some other medium". This might raise more questions than it solves regarding the long term storage and retrieval of data, so more clarification should be sought from EPA. A forth change, that simply adds an example "e.g., water, mineral oil, air" to the definition of the word "vehicle" has very little impact.

In Subpart B, relating to the Quality Assurance Unit (QAU), EPA has proposed two changes. One is that the QAU must "maintain a copy of the Master Schedule ... indexed to permit expedient retrieval". The other states that the QAU must "maintain copies of all protocols until study completion". Both of these changes were also recommended by the ACPA GLP Work Group since practically every QAU now uses a computer spreadsheet to index studies and since protocols must be archived with the original study records. Asking for expedient retrieval is no additional burden on industry, and allowing each QAU to reduce its paper storage should eventually reduce cost.

Of all the changes proposed by EPA, the one that could cause the most concern within industry is found within Subpart D on Equipment. It reads:

> "The integrity of data from computers, data processors and automated laboratory procedures involved in the collection, generation, or measurement of data shall be ensured through appropriate validation processes, maintenance procedures, disaster recovery and security measures."

Incidentally, these words were also proposed by the ACPA GLP Work Group that felt that the statement was true and concisely written. Concerns, however, arise from the interpretation of "appropriate ... measures" and how far EPA will carry this within its GLP inspection program. Some multinational companies already spend millions of dollars on these items plus computer training, and the incremental cost of placing these items under GLP could be significant (in some cases exceeding $200,000 per year if all computer systems must be validated). Perhaps an alternative would be for EPA to consider other existing standards in the computer industry or those defined within each company by their internal practices.

Going back to less controversial items, there is only one proposed change to Subpart E on Testing Facilities Operation. ACPA had hoped to eliminate the need for labeling wash bottles. However, EPA modified the suggestion to read:

> "As an alternative to labeling wash bottles and transfer bottles with the expiration date, the testing facility may develop a well-documented performance standard to ensure that the reagents or solutions have not deteriorated or are (not) outdated."

As the change is now worded, this means that the laboratory must either label the wash bottles directly or develop a performance standard which at present is unclear. Changing the words from "a well-documented performance standard" to "a standard operating procedure" would help, but overall, there is not much benefit from this proposed change.

There is some benefit from a proposed change in Subpart F that calls for test substance solubility to be determined either before the experimental start date or "concurrently according to written standard operating procedures ...". Usually, solubility testing is performed before application of a test substance. However, there are occasions when testing concurrently could shorten the study timeline and reduce costs without affecting data integrity.

Proposed changes to Subpart F also contain a couple of cautionary items. For instance, one modification states:

> "With the study director's written approval, test substance storage containers need not be retained after use, provided that full documentation of the disposition of the containers is maintained as raw data for the study."

The ACPA GLP Work Group recommended the first part of this sentence to relieve the burden of storing large numbers of containers. However, EPA added the remainder, and its description of "full documentation" is extensive. The list of items proposed for documentation is so large that it could produce more work than retaining the containers. The modification states:

> "1) (i) information of shipments pertaining to each container leaving the storage site (examples of such records are shipping request records, bills of lading, carrier bills, and monthly inventories of warehouse activity); (ii) test substance receipt records at each testing facility; (iii) complete use logs of material taken from containers; and (iv) a record of the final destination of the container, including the place and date of disposal or reclaiming, and any appropriate receipts.

2) an inventory record of empty containers before disposal, including sufficient information to uniquely identify containers, maintained in an up-to-date manner recording all arrivals of empty containers and their disposal. This record shall be maintained as raw data for this study.

3) location of facilities: where test substance is stored; where empty containers are stored prior to disposal; where records of use, shipment, and disposal of containers are maintained; and where the test substance is used in studies (i.e., testing facility)."

The other item in Subpart F that creates concern, apart from an arbitrary 12 hour interval, reads:

"Tank mixes prepared for application to soil or plants by typical agricultural practices within a 12 hour period between preparation and application, and solutions prepared for mammalian acute toxicology studies, metabolism studies, or mutagenicity studies, are exempt from requirements for concentration determinations."

While this offers some relief in terms of the testing needed on test substances, it may not go far enough since tank mixes still need to be assessed for uniformity. In cases where uniformity testing is as rigorous as concentration testing, these proposed changes may not be much help. However, it could save effort in those cases where EPA and industry can accept a visual or qualitative assessment of uniformity.

Both of the proposed changes in Subpart G are of some value to industry, since they can reduce paperwork and save effort. The first reads:

"When a reference substance for a metabolite cannot be identified prior to the beginning of a study, it is not necessary to identify the substance in the protocol."

Since one cannot identify what one does not know, many Study Directors in the past had to write protocol amendments once they knew the identity of such reference substances. Even more time consuming was the practice of writing reports on terminated studies that the following proposal would eliminate. It reads:

"Discontinued studies or studies otherwise terminated before completion shall be finalized by writing a protocol amendment providing the reason(s) for termination."

Assuming a company terminates 20 studies per year, writing one or two page protocol amendments, rather than lengthy reports, could save tens of thousands of dollars in just labor costs for the Study Director and associated personnel.

Lastly, EPA has proposed an addition in Subpart J on Reports and Records that moves it into the biotechnology realm. The change states: "For other test organisms (plants, bacteria), similarly detailed descriptions of the test system are required". The ACPA GLP Work Group agrees with the intent of this proposal but suggests that the wording be changed to: "For other test systems (plants, bacteria), similarly detailed descriptions are required."

Conclusion

The changes now being proposed by EPA represent the first opportunity to modify FIFRA GLPS since they went into effect in October 1989, and perhaps the only opportunity there will be in the next decade. While the reduction of paper, by combining FIFRA and TSCA GLPS, is the main driving force, EPA is interested in making other modest changes that could benefit the regulated industry as long as there is no negative impact to either data integrity or public safety. The ACPA GLP Work Group shares this interest with EPA and has taken this opportunity to provide comments and suggestions. When the proposed changes are published in the Federal Register later in 1998, we ask that you send your comments directly to the Public Response and Program, Resources Branch, Office of Pesticide Programs, Environmental Protection Agency, 401 M St., SW, Washington, DC 20460.

Acknowledgments

I would like to thank EPA's Office of Enforcement and Compliance Assurance for providing a draft of the proposed changes as it went to the U.S. Department of Agriculture and Congress for review. I would also like to acknowledge the following ACPA GLP Work Group Members: Ray Brinkmeyer, Dow-Elanco; Robert Brown, Agrevo; Selena Crenshaw, Griffin; Fran Dillion, Steward Ag; Tom Gale and Robert Wurz, Novartis; Clive Halder, Beyer; John Kobland, American Cyanimid; and Ray McAllister, ACPA.

Literature Cited

1. Environmental Protection Agency Office of Enforcement and Compliance Assurance, 40 CFR Part XXX, RIN-2070-AC97, Good Laboratory Practice Standards, 1997, pp 1-54, draft document.
2. ACPA GLP Working Group Recommendations, 1966, pp 1-21, available from author.
3. *Federal Register*, The Federal Insecticide, Fungicide, and Rodenticide Act, 1988.
4. *Federal Register*, Toxic Substances and Control Act, 1983.

Chapter 11

Recent Revision to the U.S. Environmental Protection Agency Residue Chemistry Guidance

Christine L. Olinger

Office of Pesticide Programs, U.S. Environmental Protection Agency, Health Effects Division (7509C), 401 M Street S.W.; Washington, DC 20460

The U.S. EPA guidance on performance of pesticide residue chemistry studies has undergone extensive revision over the past few years. In response to registrant concerns over conflicting information on residue chemistry studies, clarifying guidance was prepared for almost every type of study. These documents have been consolidated in the OPPTS 860 Residue Chemistry Guidelines. Additional draft guidance on anticipated residues and tolerances on imported commodities has been distributed since the 860 guidelines were finalized in August 1996. Highlights of all of the residue chemistry guidelines will be reviewed.

The U.S. Environmental Protection Agency (EPA) 860 Residue Chemistry Guidelines were finalized and published in August 1996. This document is a comprehensive overview of the pesticide residue chemistry studies submitted to the Agency in support of pesticide tolerances and registrations and was developed from the Subdivision O Pesticide Assessment Guidelines (Residue Chemistry) and clarifying guidance prepared in response to the reregistration rejection rate project. Highlights of the 860 guidance will be discussed in this paper. The revised Anticipated Residue and Import Tolerance guidance presented in June 1997 to the FIFRA Scientific Advisory Panel (SAP), a peer review panel external to the Agency, will be discussed as well.

The types of residue chemistry studies to be discussed are designed to answer the questions what are the potential pesticide residues in human foods and animal foods, and how much residue is present? Nature of the residue studies provide residue chemists with qualitative information on how the pesticide is altered when applied to plants or animals or ingested by livestock. Residue analytical methods are the tools used to determine how much residue is present. Storage stability studies provide information on residue stability when treated samples are stored. Magnitude of residue studies answer the question how much residue is present on the raw or processed commodity.

Rotational crop studies are used to determine appropriate intervals for planting rotated crops or whether tolerances are needed on rotated crops.

860 Residue Chemistry Guidelines

When the FIFRA amendments were first passed in 1988 requiring EPA to reregister all pesticide active ingredients initially registered prior to 1984, the Agency recognized the need to ensure that studies submitted to the Agency must be of sufficient quality to avoid endless resubmissions. The Agency examined the rejection rates of studies by discipline and identified the major factors for which they were rejected. After working with industry, residue chemists identified several areas where Agency guidance on study conduct and reporting was deficient or contradictory. Therefore, starting in 1992, the Agency developed a series of guidance documents on metabolism studies, storage stability, crop field trials, and so on. At the same time, the Health Effects Division (HED) had an on-going project re-evaluating what used to be known as Table 2, the livestock feeds table.

In 1995 the Agency initiated a project to harmonize guidelines across the Agency to avoid duplication to the extent possible. HED used this opportunity to combine all residue chemistry guidance into a single cohesive document. No longer would reviewers and registrants need to consult Subdivision O Residue Chemistry Guidelines, the Data Reporting Guidelines, the Standard Evaluation Procedures and receive conflicting information.

The draft 860 Residue Chemistry guidelines were first published in 1995. Comments received from industry, grower groups, trade associations, IR-4, etc., were addressed and incorporated into the final guidance when published in August 1996 (*1*). Since some of the guidance eventually incorporated into 860 was first developed almost five years ago and so is not new, this paper highlights the salient features of each type of study.

Nature of the Residue Studies. These studies are designed to characterize and identify the metabolites of the parent pesticide in plants and animals. Radiolabeled material is used to facilitate the identification of the total toxic residue (TTR). Dermal animal metabolism studies are required if the pesticide is to be used as a direct animal treatment, and oral studies are required if livestock feeds are treated. Assuming the dosing is correct, probably the most significant factor which led to rejection of metabolism studies was inadequate characterization and identification. Registrants and petitioners requested guidance on to what extent the breakdown products need to be characterized and identified. In response, the Agency has developed flow charts to assist investigators in determining the extent to which characterization and identification should be attempted.

The strategy for characterization and identification of extractable residues is presented in the flow chart presented in Figure 1; a second flow chart presented in Figure 2 describes procedures for characterizing non-extractable residues.

Residue Analytical Methods. Residue methods are used for enforcement of pesticide tolerances and for collection of residue data used for assessing dietary exposure and

Figure 1. Characterization and Identification of Extractable Residues

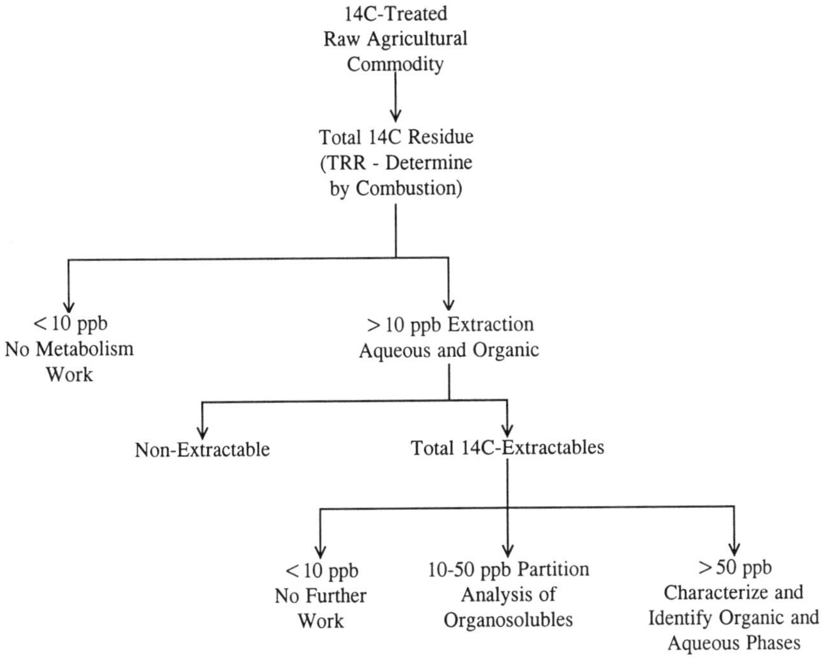

Figure 2. Characterization and Identification of Non-Extractable/Bound Residues

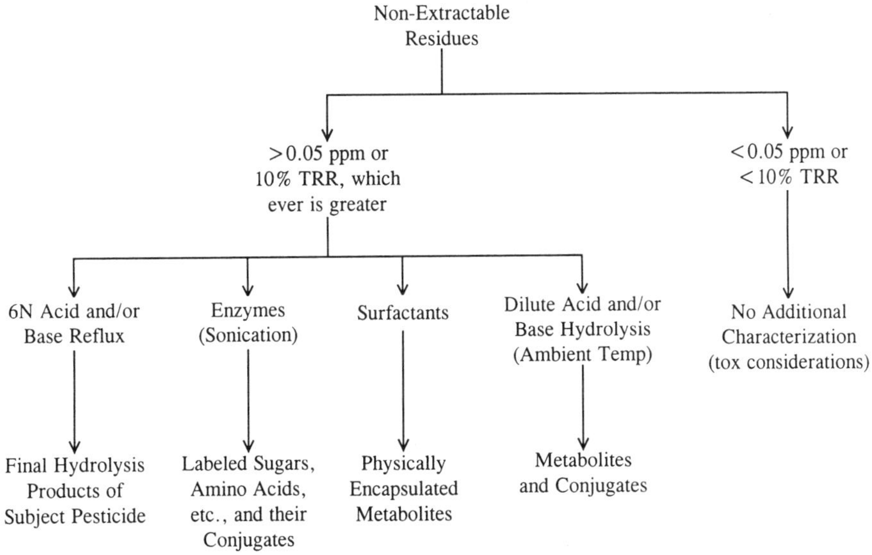

establishing tolerances. Tolerance enforcement methods are subject to specific performance criteria. The rejection rate analysis did not identify any major factors for rejection of submitted methods, but the 860 guidelines address several questions which the Agency frequently receives. Specific guidance on acceptability of immunochemistry methods, common moiety methods, and the use of internal and procedural standards is provided.

The Agency has received a few immunochemistry methods as data collection methods, but none as tolerance enforcement methods. Any immunochemistry method submitted as a tolerance enforcement method must meet the same performance criteria as any other enforcement method, such as equipment and reagents readily available to enforcement agencies and a method sufficiently specific for the pesticide and/or its regulated metabolites. Immunochemistry methods which have been submitted to date have also included a comparison to conventional methods.

The Agency frequently receives methods which involve a step which alters the chemical of interest (and some metabolites as well) to another moiety which may be present in another pesticide, also known as a common moiety method. Common moiety methods are acceptable provided a specific confirmatory method is also submitted. Should a misuse be detected using a common moiety method, an enforcement agency should have a method available to determine which pesticide may be present.

Methods using an internal standard are acceptable as long as the internal standard is added just prior to the final determinative step. Methods which employ an internal standard added toward the beginning of the method are not acceptable as this may mask unacceptable recoveries.

Independent Laboratory Validation. The Agency requires independent validation of methods to be considered for tolerance enforcement. The guidance on conducting independent validations was revised in February of 1996 and published in Pesticide Regulation Notice 96-1 (*2*). The most significant change was dropping the requirement of submitting an independent validation for enforcement methods which are significantly better than an existing method published in the U.S. Food and Drug Administration (FDA) Pesticide Analytical Manual for a given pesticide (or metabolite).

Radiovalidation. Guidance on radiovalidations is provided in the revised 860 guidelines. The Agency encourages testing treated plant and animal commodities from the nature of the residue studies through the data collection and tolerance enforcement methods. This ensures that the method(s) are capable of determining all of the total toxic residue identified in the metabolism studies.

Multi-Residue Methods. The 860 guidelines also encourage the use of the FDA Multi-residue methods as the tolerance enforcement method as an alternative to developing a single analyte method. All pesticides must be tested through the Multi-residue protocols to determine if they can be analyzed using these methods. Information on the ability of pesticides and metabolites to be analyzed using the methods is sent to FDA for incorporation in their PESTRAK database.

Storage Stability Studies. Information on the stability of pesticides and their metabolites in crop and livestock matrices must be submitted so the Agency can be assured that the conditions under which samples from metabolism and magnitude of residue studies were stored would not have led to significant degradation of the residues. These studies are usually conducted using spiked homogenates, but the Agency prefers the use of samples with field weathered residues. Clearer guidance on storage stability data requirements to support metabolism studies is provided in the 860 guidelines. Almost all of the previous guidance on storage stability studies emphasized the need for the data to support magnitude of residue studies, but information on the stability of the residue profile from metabolism studies is needed as well.

Crop Field Trials. Crop field trials are conducted to determine the maximum residue in the raw agricultural commodity at the farm gate. Residue levels found are used in estimating tolerance levels and in dietary exposure assessments. Major rejection factors identified were: i) insufficient number of trials conducted; and ii) inadequate geographical representation. In 1994 the American Crop Protection Association submitted a proposal to the Agency for determining the number and location of crop field trials to support tolerances and registrations for individual crops and crop groups. The Health Effects Division reviewed the proposal, modified it, and released a memo in 1995 instructing residue chemistry reviewers to use it when evaluating crop field trial studies. The number of trials required for each crop is based on acreage grown in the U.S. and the contribution to the U.S. diet. Previously reviewers consulted USDA statistics and specified states, not growing regions, in which the studies should be conducted. Zone maps were developed for the U.S. and the new guidance specifies zones where the trials should be located, allowing investigators to select the states in which the studies should be conducted. There is an ongoing North American Free Trade Agreement (NAFTA) project to extend the zone maps into Canada and Mexico.

Processing Studies. Processing studies are used to determine any concentration or reduction of the total toxic residue upon typical commercial processing. The revised 860 guidelines provide clearer guidance on how these studies should be conducted if residues are non-detectable in the raw agricultural commodity (RAC). With the exception of mint and citrus, if exaggerated rate field trial data are available and these studies demonstrate no quantifiable residues in the RAC, no processing studies and processed commodity tolerances are required, provided that the application rate was exaggerated by at least the highest theoretical concentration factor among all the processed commodities derived from that crop or 5x, whichever is less. A list of maximum theoretical concentration factors is included in the guidance.

Meat, Milk, Poultry, and Eggs. Magnitude of residue studies are used to assess residues in livestock commodities either from direct application to livestock or secondary residues from consumption of treated feed by livestock. When setting up a feeding study investigators must first estimate the maximum dietary burden, assuming tolerance levels and maximum percentage in the diet. The 860 guidelines include a revised livestock feeds table which includes a listing of the maximum percentage in the diet. Many animal feed items were dropped as they are no longer considered significant, including raisin waste and wet grape pomace.

Anticipated Residues

The Food Quality Protection Act of 1996 (FQPA) specifically discusses the use of anticipated residues in dietary risk assessments. When conducting dietary risk assessments EPA initially does a worst-case assessment assuming tolerance level residues and 100% crop treated. If the risk exceeds the level of concern, anticipated residues are estimated, those residues which more closely reflect the residues on food as it would be consumed. In June 1997, the Health Effects Division presented revised procedures for estimating anticipated residues for chronic dietary endpoints to the FIFRA Scientific Advisory Panel (SAP)(*3*).

Following is an outline of the procedures for estimating the anticipated residues for chronic and acute dietary risk assessments. A tiered approach is used for both, though the procedures are different for each.

Chronic Anticipated Residues. Chronic risk assessments are conducted if a chronic toxicological endpoint is identified from the toxicology studies. The Agency is concerned with the average exposure over a lifetime, since the effect is not seen unless there has been repeated exposure to the chemical of interest.

The first tier of a chronic risk assessment is essentially an upper-bound estimate whereby it is assumed that all of the commodities of interest have been treated and bear tolerance level residues. The second tier risk assessment incorporates percent crop treated information; it is assumed that if a crop has been treated, it still bears tolerance level residues. In the third tier, anticipated residues are calculated by averaging available field trial or monitoring data and incorporating percent crop treated information.

If the risk is still unacceptable the registrant may conduct special studies which would allow calculation of Tier 4 anticipated residues. Field trial and monitoring studies provide the Agency with information on the residues at the farm gate or point of distribution. Registrants may conduct further studies which would determine any reduction of residues during typical processing by the consumer or the seller. Cooking, washing, and market basket studies are often conducted to demonstrate that for many pesticides, the residues determined at the farm gate exceed that to which the consumer is exposed. These are expensive studies, so they are not usually done unless the registrant has a great incentive to do so.

Acute Anticipated Residues. Acute dietary risk assessments are done less frequently than chronic assessments. They are conducted if the toxicologists identify an acute hazard, which is defined as an adverse effect from exposure to a pesticide within a single day. The most frequent endpoints for which acute risk assessments are conducted are acute cholinesterase inhibition and developmental effects, where a single exposure during gestation could affect the health of the developing fetus. Because the Agency is concerned with a single exposure, the maximum exposure must be evaluated, not the average.

Tier 1 for acute dietary risk assessments is the same as for chronic assessments, assuming tolerance level residues and 100% crop treated. For tier 2, there is still an assumption of 100% crop treated and tolerance level residues for unblended commodities, but residues are averaged for highly blended commodities such as oil, grape juice, and sugar, since it is highly likely that treated commodities will be blended with untreated commodities.

Tier 2 is also an over estimate, since it assumes all non-blended commodities bear tolerance level residues, when actually there is likely a range of residues throughout the food supply. Tier 3 dietary risk analyses employ the use of Monte Carlo techniques to estimate the risk, assuming a range of food consumption and residue values from field trial data (*4*). Monte Carlo analyses provide a probability of the highest consumers will actually ingest the commodities bearing the highest residues. Should the risk from a Monte Carlo analysis still exceed the Agency's level of concern, the registrant may be required to go on to Tier 4 and conduct market basket surveys. These are very expensive studies, but they give the Agency the best information on residues in foods as purchased by the consumer.

In the final report by the FIFRA Scientific Advisory Panel following the presentation in June 1997, the SAP expressed concern over the food consumption data used by the Agency, the 1977-78 USDA food consumption survey. Previously, the Agency has attempted to adopt more recent surveys, but they have not been able to since they were not designed for the OPP Dietary Risk Evaluation System. However, EPA is currently in the process of evaluating the most recent USDA Food Consumption survey and hopes to incorporate it before the end of 1998.

Import Tolerances

EPA has the authority to set pesticide tolerances under Sec. 408 of the Federal Food, Drug, and Cosmetic Act. Most of these tolerances are set in conjunction with the registration of the pesticide on the subject crop. Increasingly EPA has been petitioned to establish tolerances in the absence of a U.S. registration. These are commonly referred to as import tolerances because the pesticide could not be legally used on the subject commodity in the U.S., and, therefore, only imported commodities would bear legal residues of the pesticide. It is important to note that there is no legal distinction for "domestic" and "import" tolerances. The term "import tolerance" is a term of convenience used to denote tolerances in the absence of a U.S. registration.

In April 1997, the Agency released a draft document describing the data requirements for import tolerance petitions (*5*). While most data requirements are generally the same for tolerances in/on imported commodities (as for tolerances with a U.S. registration), there are some differences since there is not a need for data which support registration, and the location of some studies may need to be different. Summarized below are the highlights of the import tolerance guidance.

Product Chemistry Data Requirements. The Agency needs some basic information on the pesticide chemical itself. The petitioner must submit product chemistry information, such as, how the technical material is manufactured and what the expected impurities are (*6*). Some basic physical and chemical characteristics of the pure or technical material must also be described. The inert ingredients in the formulation must be identified as the Agency may have some concern about the potential dietary risk from these materials. Fewer studies are required for an import tolerance since many product chemistry studies are used to support the registration of a product and not a tolerance (associated with the use of the product).

Toxicology Data Requirements. Fewer toxicology studies are also required to support an import tolerance than those needed for a product registration. Acute studies used for precautionary statement labeling are not needed because these are required for a product registration. Dermal and inhalation toxicology studies are not needed because these are used only for risk assessments associated with occupational and residential exposure. Only those studies needed for hazard identification and dose response determination associated with acute and chronic dietary risk assessments are required.

Residue Chemistry Data Requirements. Most residue chemistry studies which would be required for a tolerance with a U.S. registration will be required for a tolerance on imported commodities as well. Some livestock and processing studies may be waived, as described below. The most critical part of the residue chemistry data requirements is the procedure for determining the number and location of crop field trials.

Normally, livestock metabolism studies are required if the pesticide will be applied directly to animals or if it will be applied to a crop which may be fed to livestock. However, for import tolerances, petitioners may be able to propose waiving livestock studies for those crops which would normally trigger such studies. If the countries exporting the commodity (for which a tolerance is sought) do not have significant export of livestock commodities into the U.S., and the petitioner can convincingly demonstrate that the commodity will not likely be used as an animal feed in the U.S., then livestock studies will not be required. This is also true for processed commodities used for livestock feeds as well.

Table 1 of 860.1000 lists the raw agricultural commodities (RAC) for which processing studies are required (*7*). Import tolerance petitioners may be able to demonstrate that some processing studies may be waived if the country in which the pesticide is used does not have significant export of the processed commodity or the RAC is not likely to be processed in the U.S. In other cases, a processing study may be required, but a tolerance may not be needed on the processed commodity, even if significant concentration is observed. An example of this situation would be wet apple pomace. Wet apple pomace is not imported into the U.S., but meat from livestock that have consumed the pomace in a foreign country may be imported, so information on the residue levels is necessary.

Number and Location of Foreign Crop Field Trials. Previously, the Agency has had no written guidance on the number and location of crop field trials needed for import tolerances. The proposed method for determining the minimum number of field trials is based on the number of trials required for a U.S. registration, the percent of the commodity available for consumption (in any form) which is imported, and the percent of the commodity consumed in the U.S. diet. The first step is for the petitioner to determine the percent imported into the U.S. from those countries in which the petitioner markets or intends to market the pesticide. Calculation of the percent imported must include all forms of the commodity, such as fresh fruit, juice, wine, juice concentrate, dried fruit, and so on.

If the percent imported is less than 75%, the petitioner uses Table I. Using another table which lists the number of field trials required for a U.S. registration, available in the 860 guidance, and the percent imported, the petitioner can easily find on this table the minimum number of trials required (*8*).

Table I. Number of Field Trials Required for an Import Tolerance (Less than 75% Imported into U.S.)

Required No. of Field Trials for a U.S. Registration	Number of Field Trials Required for an Import Tolerance		
	0 - 10%[1]	10 - 35%[1]	35 - 75%[1]
20	5	16	20
16,15	5	12	16
12	3	8	12
9,8	3	5	8
6,5	3	3	5
3	2	3	3

[1]Percentage of commodity imported into U.S. available for consumption (weight basis).

For commodities in which most of what is available for consumption in the U.S. is imported, i.e. greater than 75%, the petitioner refers to Table II. In this case the minimum number of trials required is based on the amount in the U.S. diet. This information is available in the residue chemistry guidance (*8*).

Table II. Number of Field Trials Required for an Import Tolerance (Greater than 75% Imported into U.S.)

Maximum Percent in U.S. Diet	No. of Trials Required
0 - 0.05	3
0.05 - 0.2	8
0.2 - 1.0	12
>1.0	16

The next question to be answered is in which countries should these trials be conducted? All countries in which the pesticide is marketed or intended to be marketed must be represented. Usually only data would be required from those countries which represent at least 5% of the total imports. The number of field trials in each country should be relatively proportional to the amount imported into the U.S.

The aforementioned tables refer to the 'minimum number of field trials'. The number of field trials may need to be increased to ensure all formulation classes be adequately tested in side-by-side trials, and that all countries be adequately represented. The total number of field trials for those commodities which require more than eight trials may be reduced by 25% if residues in the commodities are all below the limit of quantitation, and the crops are not being used as representative commodities to obtain crop group tolerances (*9*).

Fewer than three trials may be conducted if the dietary consumption is very low, **and** a relatively small amount of the commodity is imported into the U.S. Four independent samples must be collected from each test plot if less than three trials are conducted. Petitioners should either consult OPPTS Guideline 860.1500 or contact the Agency before proceeding if they believe that fewer trials are warranted (*8*).

In the Agency reregistration program, some pesticide registrants have requested dropping the U.S. use on a commodity, but retaining an existing tolerance to cover imported commodities. If U.S. crop field trial studies are available, but foreign data are not, petitioners frequently ask whether U.S. data may be substituted for foreign data. If the petitioner can adequately demonstrate the U.S. data are representative of the foreign region of interest with respect to climate and agricultural practices, then U.S. data may be substituted for up to half of the trials. A minimum of three trials must be conducted outside the U.S. For U.S. registrations there are 10 regions from which data are required; the Agency has plans to extend those regions into Mexico and Canada as part of the pesticide registration harmonization process under NAFTA.

Codex Considerations. The Agency is frequently asked why bother requiring data at all if some other country or international organization has set a tolerance or maximum residue limit (MRL) on the pesticide/commodity combination of interest? To address these concerns, the Agency will consider petitions with limited review of the residue chemistry data under certain low-exposure/risk situations. At this time all toxicology data would still be required, but the possibility of sharing reviews is under consideration.

If a Codex MRL has been established, then the petitioner may submit a petition for limited review of the residue chemistry data if it meets the certain conditions. The commodity must not contribute significantly to the U.S. diet, and the pesticide should not be already considered a human dietary risk concern. An enforcement residue analytical method should be submitted if one is not available in the FDA Pesticide Analytical Manual. Codex must regulate the same metabolites as the U.S. considers to be of concern. A risk assessment would be conducted using the Codex MRL; provided the risk is acceptable, the Codex MRL would be established as the U.S. tolerance.

It is the intention to release the draft incorporating the comments from the SAP in mid-1998. This will be considered interim guidance. The NAFTA Technical Working Group on Pesticides will be developing import tolerance guidance, and it is the intention of the U.S. to incorporate the final NAFTA document into the Agency guidelines to the extent possible.

Conclusion

Highlights of the most recent revisions to the U.S. Residue Chemistry guidance have been provided. The 860 guidelines are a comprehensive document which should be used when conducting or evaluating studies. The Anticipated Residue Guidance and Import Tolerance Guidance were both presented to the SAP in June 1997 and revisions to the draft documents incorporating the recommendations of the SAP will be released before the end of 1998.

Acknowledgments

Ms. Olinger would like to recognize the assistance of all the residue chemistry reviewers in the Health Effects Division, particularly Richard Loranger, Randolph Perfetti, Susan Hummel, Fred Ives, George Kramer, Bernard Schneider, and Christina Swartz.

Literature Cited

(1) OPPTS Test Guidelines, Series 860, Residue Chemistry (August 1996).
(2) PR Notice 96-1, "Tolerance Enforcement Methods - Independent Laboratory Validation by Petitioner", February 7, 1996.
(3) Anticipated Residue Guidance - Presentation to the FIFRA Scientific Advisory Panel (Draft - April 1997)
(4) "Guiding Principles for Monte Carlo Analysis" U.S. EPA Publication; document No. EPA/630/R-97/001; (March 1997)
(5) Draft Guidance on Import Tolerances (October 1997 - in preparation)
(6) OPPTS Test Guidelines, Series 830, Product Chemistry (August 1996).
(7) OPPTS Test Guidelines, 860.1000, "Background" (August 1996).
(8) OPPTS Test Guidelines, 860.1500, "Crop Field Trials" (August 1996).
(9) OPPTS Test Guidelines, 860.1340, "Residue Analytical Methods" (August 1996).

OPPTS Test Guidelines can be accessed from the Government Printing Office Bulletin Board at http://fedbbs.access.gpo.gov/libs/epa_860.htm (or 830.htm). All other documents may be found by searching the EPA Home Page at http://www.epa.gov.

Chapter 12

The Effects of the Harmonization of Regulatory Requirements on the Crop Protection Industry

Frederick L. Groya

American Cyanamid Company, Clarksville and Quacherbridge Roads, Princeton, NJ 08543

Pesticide regulations are currently being harmonized among regulatory agencies. Areas being harmonized include data requirements, study guidelines, data formatting, maximum residue limit setting, and risk assessment procedures. These efforts at harmonization are driven by international trade issues, decreasing government resources and a desire to bring world-wide standards to a common level. The effect that harmonization will have on the crop protection industry will depend on the direction and the extent to which it proceeds. Negative effects will result if standards are based on policies of pesticide use reduction, are a compilation of all existing requirements, or involve the inappropriate transfer of environmental risk assessments. Positive effects will result if harmonization proceeds to the point where a core set of data requirements, study guidelines and assessment procedures based on sound scientific principles are widely adopted. Potential benefits could consist of a more efficient planning process, lower development costs and earlier market entry.

The regulation of pesticides today is conducted, for the most part, on a national basis with each country having its own set of regulations and registration criteria. The objective of these national registration systems is to confirm that a pesticide is safe to humans, non-target organisms and the environment when used according to label directions. The core set of requirements that must be met for registration are similar from country to country. Technical experts in the fields of toxicology, ecotoxicology, metabolism, residue chemistry and environmental fate have been able to agree, for the most part, on a core set of requirements based on accepted scientific principles. Concurrence among the experts has not been possible in all areas, however, leading to some differences in data generation methods and data requirements. Furthermore, when establishing official regulatory policy, governments must take into account national values and public opinion which can vary greatly from one country to another. The lack of complete agreement on scientific issues and the diversity of national values has led to sufficient variation in regulatory requirements to create difficulties and inefficiencies for the crop protection industry when registering pesticides in more than one country.

Today, there is an effort on several fronts to harmonize pesticide regulations among government regulatory agencies. That is, there is an effort to find a common approach to how the potential human health and environmental risks associated with the use of pesticides are identified, characterized and managed. The purpose of this paper is to explore what is currently being done to harmonize the regulation of pesticides and to predict the effects it will have on the crop protection industry.

Types of Regulations

The regulation of pesticides is a complex exercise in measuring the inherent hazard a product may present to humans and the environment, the exposure to that product that is expected to occur during normal use, and the risk that will result from the use of that product. Therefore, before the harmonization of regulatory requirements can be discussed, the various aspects of these requirements must first be defined.

The term "data requirements" refers to the list of data that must be generated and submitted to the regulatory authority. Data are usually required in the areas of product chemistry, residue chemistry, metabolism, mammalian toxicity, aquatic toxicity, avian toxicity, effects on non-target organisms and environmental fate. These data will provide information on the inherent hazard of the product and the amount to which humans, wildlife and the environment will be exposed. Today, between 150 and 200 studies must be conducted at an expense of US$50-100 million to fulfill the necessary requirements for the full registration of a major pesticide.

The study guidelines or protocols are the methods or procedures by which a particular study must be conducted. Even though many countries require the same data to be generated, the way in which the data are to be generated is not always the same. Some regulatory agencies have provided very detailed guidelines on how a particular study should be conducted. Failure to conduct the study according to these guidelines will render the data unacceptable to the regulatory authorities. It is possible, therefore, to conduct a study that is acceptable to one government agency but not to another. The area of study guidelines provides much opportunity for harmonization.

Formatting refers to how the data and other required information must be packaged together for submission to a regulatory authority. As with the data requirements and the study guidelines, many government agencies have prescribed unique, detailed guidelines on how to submit the data. This requires that the same information be re-packaged by the registrant for each country to which it is being submitted.

A maximum residue limit (MRL) or a tolerance is the level of residue that can legally remain in or on a raw or processed food commodity. MRLs are set based on the results of actual field residue studies. The inherent toxicity of the product also is considered to ensure that the level of dietary exposure to that product will not result in undue risk. Countries set their own MRLs based on local residue trials and often use different criteria to set the MRLs. Some countries use the highest residue observed in field trials while others will use the mean or median value. It is common for MRLs to differ from country to country.

When deciding whether a pesticide should be registered for use, a regulatory agency must assess the risk that the use of that product will pose to humans and the environment. Since risk is a function of both hazard and exposure both factors must be considered as part of the assessment. For example, a product with a high level of mammalian toxicity will present a negligible risk to humans if exposure to the product is sufficiently low. Furthermore, a product to which there is a high level of exposure such as through residues in food and water, could also present a negligible risk if it is relatively non-toxic.

Because risk assessment is the most politicized aspect of the regulatory process, there can be large differences in how assessments are done from one country to another. An example of this is the Food Quality Protection Act enacted into law in the United States in 1996 (*1*). This law established new risk assessment procedures that are very different from those used any where else in the world. This trend against harmonization could result in the loss of registrations for certain products in the United States while registrations are maintained elsewhere.

Once the risk is characterized, regulators have to decide what level of risk is acceptable. Decision making schemes are usually a matter of public policy rather than a scientific determination. In the United States, for example, one in one million or 10^{-6} has generally been considered acceptable cancer risk for consumers exposed to pesticide residues in foods. In other countries, decisions will be based on a "precautionary principle" instead of a risk-based approach which employ "cut off" criteria. In these schemes, exposure and hazard are not considered together and a product can be determined to be unacceptable simply based on a certain hazard level or an exposure level. For example, an exposure level of 0.1 ppb in ground water could render a product unregistrable in the European Union regardless of its toxicity or hazard. Differences in decision making schemes can result in a product being registered for use in one country while being banned from use in another even though conditions of use in the two countries are similar. An example of this is the herbicide, atrazine, which is banned or severely restricted in certain European countries but widely used in a safe manner in the United States.

The general policies of national governments differ greatly from one another and affect the registrability of products. These differences can be attributed to differences in public perception of pesticides that, in turn, shape government policy including data requirements. If public perception is that pesticides are bad, then government policy and the resulting regulations will reflect this perception. If public perception is that pesticides are useful and necessary agricultural tools when used properly, then government policy and regulations will reflect that perception. In other words, public perception will set the tone for national regulations.

Because perceptions vary greatly from one region to another, general policy, and thus pesticide regulations, naturally tend to differ from country to country. Differences in public policy on environmental issues between the European Union and the United States have resulted in differences in the requirements for environmental fate and ecotoxicological studies. Another example can be found in Japan, where the requirements in the area of animal metabolism are much more extensive than those in most other countries.

Reasons for Harmonization

There are three reasons that can be identified as the driving forces behind the effort to harmonize the regulation of pesticides: (1) international trade, (2) increased efficiency and (3) a desire to bring world-wide standards to a common level. It is important to note that although industry will benefit by certain aspects of harmonization, it is governments that must see a benefit in it for themselves in order for harmonization to proceed. How public policy and, thus, government interests are affected will determine if, and to what extent, harmonization will occur.

The most important, and earliest, force behind the harmonization of pesticide regulations has been economic as exhibited by international trade issues. Differences in the regulatory schemes between countries have restricted the movement of pesticides and treated commodities and were viewed as non-tariff trade barriers. Therefore, the harmonization of pesticide regulations, in particular the harmonization of MRLs, was first undertaken with the objective of ensuring the free flow of goods across international borders. The formation of regional trade groups such as the European Union and the North America Free Trade Agreement have provided the structure and mechanisms for countries to pursue harmonization in order to achieve the economic objectives of the member countries.

Another factor that has promoted efforts toward the harmonization of pesticide regulations has been the desire to make the registration process more efficient thus conserving limited government resources. On a world-wide basis, the registration of a plant protection product is very redundant and resource intensive as each individual government repeats the assessment process. Efficiencies could be gained by limiting the amount of information that would need to be reviewed by each individual country. This could be achieved if all national governments could rely on one single expert review of the information that was not country-specific.

A third force behind harmonization is a desire on the part of some governments to extend their policies, regulations or risk assessment procedures to other world areas. There is a desire to promote public health and environmental protection on a global scale and to ensure that all regulatory systems are based on scientific and ethical principles. Specific reasons that a country might have for extending its regulations and procedures beyond its borders include (1) a concern for the types of products used and the risk that these may pose on foreign populations or environments, and (2) a concern for residues that may be found on foreign-grown commodities that it may import. The converse of the reasoning regarding trade actually comes into play in this area. Instead of reviewing regulations to ensure that trade barriers are removed; trade agreements are reviewed to ensure that environmental standards are not compromised. Whatever the reasoning, many developed, industrialized countries with sophisticated regulatory agencies do, to some degree, deal with pesticide use outside of their borders.

Harmonization Efforts

Efforts to harmonize pesticide regulations are currently under way on both the regional and international levels. This section will briefly describe these efforts.

International Efforts. One of the first attempts at harmonization was the establishment of the Codex Alimentarius in the 1960s under the World Health Organization (WHO) and the Food and Agriculture Organization (FAO). The objective of Codex is to establish MRLs for the residues of pesticides that could be found on commodities that are traded internationally. The Codex MRL provides a mechanism to alleviate trade issues that could occur related to pesticide usage and is recognized by the World Trade Organization for this purpose. Furthermore, some less-developed countries will rely on the Codex MRL in lieu of their own MRL setting capabilities. Unfortunately, Codex MRLs are not recognized by all governments due to disagreements over MRL-setting procedures and the relevance of the MRL to its citizens' diets.

Other examples of the United Nations involvement in pesticide use are (1) the WHO Recommended Classification of Pesticides by Hazard (*2*), (2) the FAO Pesticide Specifications (*3*), (3) the FAO Code of Conduct for the Distribution and Use of Pesticides, and (4) the FAO/UNEP Joint Program on Prior Informed Consent (PIC).

The Organization for Economic Cooperation and Development (OECD), formed in 1961, is an organization of the governments of the industrialized democracies. The primary objective of the OECD is to allow its member countries to adopt their domestic policies in such a way as to minimize conflict with other member countries. As recently as 1992, the OECD began work on pesticide issues in order to reduce differences in national systems that result in redundancies in the regulatory process (*4*). Five specific areas in which the OECD is active include (1) test guidelines, (2) data requirements, (3) hazard assessment, (4) reregistration of older pesticides and (5) risk reduction.

Regional Efforts. With the increase in the formation of regional trading blocks around the world, there has been a corresponding increase in regional efforts to harmonize pesticide regulations. The extent to which harmonization has occurred will mirror the degree of organization of the trading block.

One of the more advanced trading groups is the European Union (EU). A high level of organization has allowed the EU to advance considerably in the harmonization effort. The registration process in the EU is being harmonized under Directive 91/414 which has been in effect since mid-1993 (*5*). Under the new directive, all member states have a common set of data requirements, study guidelines and decision-making criteria. However, even in this highly "harmonized" system there is no EU-wide registration. Only an active ingredient is authorized on an EU basis; thereafter, the end-use products still must be registered by each individual member state. Even the EU active ingredient authorization is a consultative process which can be influenced by the policies and political agendas of the individual member states.

Under the North America Free Trade Agreement (NAFTA), certain aspects of the pesticide registration process are being addressed by the United States, Canada and Mexico. The most progress has occurred between the US and Canada since efforts have been ongoing for a longer period of time under the NAFTA precursor, the Canada-US Trade Agreement. Areas being addressed by the NAFTA Technical Working Group on Pesticides include (1) the resolution of "trade irritants" caused by differing or missing MRLs; (2) harmonization of data requirements, test guidelines and risk assessment procedures; and (3) improving operating efficiencies by sharing work. The work being done under NAFTA is a good example of how regulatory agencies are using a structure that was put in place to facilitate trade in order to gain regulatory efficiencies and save limited resources.

Another regional trading block that has formed in the Western hemisphere is the Mercado Comun del Sur (MERCOSUR) which is comprised of Argentina, Paraguay, Uruguay and Brazil. MERCOSUR member states are still in the process of negotiating changes in their national regulations that will result in one harmonized registration process for the group. As a first step, a positive list of active ingredients was established to allow for the free movement of products within the region. The completion of a harmonized pesticide registration system for the trading block is expected by the year 2000.

Other efforts to harmonize pesticide regulations can be found in West Africa with the *Comite Permanent Interetats de Lutte Contre la Secheresse dans le Sahel* (CILSS). Burkina Faso, Chad, Gambia, Guinea Bissau, Mali, Mauritania, Niger and Senegal have set up a central regulatory authority that will review and grant registrations on behalf of each member country. Another example can be found in Central America where the countries of that region have established a common system for pesticide product labeling.

Bilateral Efforts. In addition to the efforts to directly harmonize pesticide regulations, there are also numerous examples of bilateral efforts between countries that will indirectly result in a certain degree of harmonization. These bilateral efforts occur as (1) projects that focus directly on the registration system of a developing country, (2) projects that focus on pesticide policy development, in general, and (3) pesticide use stipulations attached to aid programs. An example of the first type of bilateral effort is the US EPA's project to improve the regulation of pesticides in Indonesia (*6*). In this project, the US EPA is working directly with the Indonesian Ministry of Agriculture.

Northern European governments provide examples of attempts to harmonize pesticide use policies. The German Agency for Technical Aid (GTZ)/University of

Hanover Pesticide Policy Project that is funded by the German Ministry of Economic Cooperation and Development (BMZ) targets pesticide use in developing countries. One example is the involvement of the GTZ in the pesticide risk reduction program in Thailand (*7*). Also, the Swedish International Development Cooperation Agency (SIDA) has been involved in pesticide use policy in Central America.

Developed countries can also indirectly affect the pesticide use policy of a lesser-developed country via financial aid programs. Oftentimes, aid for agricultural programs will have a stipulation that certain pesticides cannot be used or can only be used to a limited extent. In essence, this is resulting in a certain level of harmonization of pesticide use policy, albeit on a temporary basis. The FAO has established guidelines for pesticide procurement to ensure that hazardous materials are used only in an appropriate manner that minimizes any risk resulting from the use of the product. These guidelines have been established in recognition of the fact that many developing countries that receive pesticides by way of donations do not have internal mechanisms to control the use of those pesticides. The World Bank has a similar policy to the FAO guidelines for agricultural development programs that involve the use of pesticides.

Harmonization by Default. To complete the picture, mention must be given to a type of harmonization that occurs by default. Many lesser-developed countries rely directly on the regulatory actions of developed countries. In these cases, a country that does not have an extensive national registration system will rely on documented proof that a product is registered in a developed, industrialized country. Therefore, by default, the pesticide registration system of the developed country becomes the system of the developing country.

Effects of Harmonization

Different groups have different opinions on the ultimate effect that harmonization will have on the regulation of pesticides. Some governments see it as an opportunity to upgrade the regulatory systems of other countries as well as a means to achieve efficiencies within their own systems. Some public sectors fear that harmonization could result in the lowering of standards as governments seek a common, lower ground. Other sectors fear an opposite development toward the highest common factor making the regulatory process more difficult and expensive. Harmonization could have either positive or negative effects on the crop protection industry depending on the direction and the extent to which it proceeds.

Negative Effects. The ultimate determinant of the effect harmonization will have on the crop protection industry will be whether or not sound, scientific principles are maintained as standard requirements, guidelines and assessment procedures are widely adopted by national governments. If these standards are not based on sound, scientific principles but, instead, are based on policies of pesticide use reduction, then harmonization will have a very negative effect on the industry. The regulations that are enacted by national governments today to keep useful crop protection products off the market would have devastating effects if these regulations were to be adopted on a global basis. The regulatory process would become longer and more costly with fewer products being brought to the market.

Harmonization will also have a negative effect if adopted standards are simply a compilation of each country's requirements. What could result is a situation similar to that in the EU where data requirements are almost a compilation of the most stringent requirements maintained by the member states. We have seen that governments are not always willing to let go of their most cherished regulations and policies regarding pesticides. A simple compilation of the most stringent requirements will negatively affect the industry by increasing development costs.

Complete harmonization to the point of conducting global risk assessments would be impractical and not in the interest of the industry. Because conditions vary from one area to another, and in the absence of good predictive models, a certain amount of country-to-country evaluation and assessment will always be necessary. Therefore, it would not always be appropriate to transfer the results of an environmental risk assessment from one country to another. Furthermore, the area of risk management, or decision-making, is where politics will most likely come into play and decision-making criteria are oftentimes based more on public opinion than on sound science. It is better to not harmonize this final step in the registration process than to adopt unreasonable decision schemes.

Positive Effects. Positive effects on the crop protection industry due to harmonization will only occur if the harmonized standards are based on practical and sound scientific principles. Therefore, the potential positive effects presented here are done so with this stipulation in mind. The positive effects of harmonization on industry could primarily occur in the form of a more efficient planning process, lower development costs and, most importantly, earlier market entry. Some benefits will be dependent upon national governments' willingness to rely on one evaluation of the data as opposed to each government individually repeating the evaluation.

More Efficient Planning. A common set of data requirements and study guidelines would create a more efficient planning process for industry. Today, a considerable amount of time and energy is spent assuring that the multitude of national requirements and varying guidelines are properly reflected in a development program. For example, for a major new pesticide that will be developed for use in the major agricultural markets of Europe, North America, South America and Asia, the regulatory requirements of the EU, the United States, Canada, Brazil, Argentina, Australia and Japan must be studied carefully, at a minimum, to ensure that no data requirements are missed. Also, each study protocol must be written to assure that the resulting data will be accepted by each regulatory authority. Under a harmonized system, each of these countries would have an identical set of requirements and guidelines making the planning process much more straight forward. Also, standard protocols could be used consistently as opposed to re-writing protocols each time to assure that all aspects of each government's requirements are addressed. Less time would also be needed to track the changes that are always occurring to individual national regulations. The man-power resources currently expended on these planning activities could be redirected into the R&D effort.

Savings in man power resources could also occur after the data are submitted to the authorities. Today, the technical experts in R&D companies must spend an inordinate amount of time preparing information for and meeting with the regulators of the different countries where registrations are being sought. Under a system of a single primary evaluation, much of this repetition could be eliminated.

One common guideline for the formatting of dossiers will also result in a more efficient development operation. With a common guideline, the core data could be formatted and summarized only one time. The core package could then be modified to address any unique characteristics of an individual country.

Lower Development Costs. Lower development costs would be realized by the industry if requirements and guidelines are rationalized in a practical manner. This assumes that studies that are done specifically for only one or two countries today will be eliminated as all countries agree on a single set of studies and protocols. A compilation of each country's existing set of requirements would not result in any benefits in cost reduction. Also, certain studies could be done less expensively if a single protocol has been established instead of doing more expensive studies that try to address several differing protocols.

Other cost savings would be realized if governments could harmonize the MRL setting process. If this is done properly, governments could rely on studies done under comparable conditions instead of insisting that all residue trials be done locally. The EU is already approaching this by dividing the European continent into northern and southern zones for the purposes of conducting residue and efficacy trials. Another benefit of MRL harmonization would be the removal of the need to apply for import tolerances as is done in the United States. Today, tolerances or MRLs must be applied for in the US if imported commodities contain pesticide residues. This requires a full evaluation and the expenditure of resources by the US authorities. Under a harmonized system, the US could simply accept the MRLs established by the countries where the product is used and forego the expense of another complete evaluation of the data.

One benefit that should not be expected from harmonization is a shortening of the period of time required for data generation. There will always be the need to conduct long-term studies such as the mammalian chronic/oncogenicity studies so that the overall timelines would not be reduced.

Earlier Market Entry. The ultimate potential benefit for industry that could result from harmonization is earlier market entry in a greater number of countries than is possible today. Theoretically, faster entry into the market place in multiple countries could occur if all countries relied upon one evaluation of the core data. This reliance upon one evaluation would reduce the work load of regulatory agencies around the world who would not have to repeat the primary evaluation. This reduced workload would allow for a quicker review by the agency doing the primary evaluation and by the other national governments who would then review only the country-specific aspects of the product.

Conclusion

The harmonization of regulatory requirements is seriously under way on several bilateral, regional and international levels. This effort will continue since it is being driven by government regulatory agencies. This process could have both positive and negative effects on the crop protection industry depending on the direction and the extent to which it occurs. Harmonization could have a very negative effect if the adopted standards are based on policies of pesticide use reduction, are a compilation of all existing requirements or involves the inappropriate transfer of environmental risk assessments. On the other hand, the industry could benefit significantly if harmonization proceeds to the point where a core set of data requirements, study guidelines and assessment procedures based on sound principles are widely adopted. Potential benefits could consist of a more efficient planning process, lower development costs and earlier market entry.

Literature Cited

1. Food Quality Protection Act of 1996. Public Law 104-170. August 3, 1996.
2. WHO Recommended Classification of Pesticides by Hazard and Guidelines to Classification 1996-1997. International Programme on Chemical Safety. WHO/PCS/96.3.
3. Manual on the Development and Use of FAO Specifications for Plant Protection Products. FAO Plant Production and Protection Paper No. 85. Rome 1987.
4. Grandy, N. J.; Richards, J. *Brighton Crop Protection Conference.* 1994, pp.1409-1418.
5. Council Directive 91/414/EEC. July 15, 1991. Official Journal of the European Communities No. L230/1-32.
6. Office of Pesticide Programs Annual Report for FY 1997. US Environmental Protection Agency. EPA735-R-97-003. January, 1998. p.23.
7. Jungbluth, F. *Crop Protection Policy in Thailand.* Pesticide Policy Project Publication Series No. 5. University of Hannover. December 1996.

Chapter 13

Implementation of the Food Quality Protection Act of 1996

Margaret J. Stasikowski and Kathleen A. Martin

Health Effects Division (7509C), Office of Pesticide Programs, U.S. Environmental Protection Agency, 401 M Street S.W., Washington, DC 20460

The Food Quality Protection Act of 1996 amended both laws under which pesticides are regulated in the United States. Major provisions of the Federal Insecticide Fungicide and Rodenticide Act include a 15-year registration renewal cycle, acceleration of the registration process for 'safer' pesticides, and the establishment of separate programs for minor uses and antimicrobials. Major provisions of the Federal Food Drug and Cosmetic Act include a single, health-based standard for residues on all foods, special considerations for infants and children, a schedule for reevaluation of all existing tolerances, testing for endocrine disruptors, uniformity of tolerances, and enhanced enforcement capabilities for the Food and Drug Administration. How the Agency has been implementing the new law will be described, with particular emphasis on risk assessment and science policy aspects.

The Food Quality Protection Act of 1996 (FQPA) is a law that was enacted by the President on August 3, 1996, *(1)* to enhance the safety of the American food supply. It is the most significant piece of pesticide and food safety legislation enacted in many years. In fact, FQPA is the first major revision of pesticide laws in over 30 years. This law is so important because it reflects a national commitment to greater protection of infants and children from the possible effects of pesticide residues in food. FQPA is also important because it introduces a good measure of common sense into the process of pesticide regulation by recognizing that the state-of-the-science is constantly evolving. Two good examples of this are the elimination of the Delaney Clause *(2)* and the provision that establishes a system for periodic review of all pesticide registrations. The Delaney Clause is the old provision where raw and some processed foods were held to

Disclaimer. The views presented in this paper are those of the authors and not necessarily those of the U.S. Environmental Protection Agency (EPA or the Agency).

different standards of safety. In drafting FQPA, Congress was able to abolish it (for pesticides) because our understanding of toxicity is a great deal better than it was in the 1950s, which is when the Clause was enacted.

Much work and thought preceded the enactment of FQPA. Over the past several years, the scientific and public health communities have been raising concerns about the unique susceptibilities of infants and children to environmental toxins. In 1988, The U.S. Congress asked the National Academy of Sciences to evaluate the Agency's existing risk assessment practices to determine whether or not they adequately considered the unique potential for risks to infants and children. In 1993, the Academy published its report, *Pesticides in the Diets of Infants and Children*, concluding that infants and children may have significantly different exposures and/or responses to pesticides than adults *(3)*. Because of these and other differences, the National Academy of Sciences recommended, among other things, that: "Because there exist specific periods of vulnerability during postnatal development, the committee recommends that an uncertainty factor up to the 10-fold factor traditionally used by EPA and FDA for fetal developmental toxicity should also be considered when there is evidence of postnatal developmental toxicity and when data from toxicity testing relative to children are incomplete.....in the absence of data to the contrary, there should be a presumption of greater toxicity to infants and children" *(4)*. Eventually, this recommendation, among others, found its way into the provisions of FQPA.

Major Provisions of FQPA

EPA's Office of Pesticide Programs regulates pesticides under two statutes: FIFRA (The Federal Insecticide, Fungicide, and Rodenticide Act) and FFDCA (The Federal Food, Drug, and Cosmetic Act). Generally speaking, most of the regulatory aspects of the pesticide program are governed by the provisions of FIFRA while the human health risk assessment aspects fall under the scope of FFDCA. FIFRA requires that pesticides be 'registered' (licensed) by EPA before they are sold or distributed for use in the United States. FFDCA authorizes EPA to set tolerances, or maximum legal limits, for pesticide residues in food. FQPA amended both of these statutes.

On the FIFRA side, FQPA requires that the Office of Pesticide Programs, among other responsibilities: review its pesticide registrations periodically to ensure that they still meet the most up-to-date science; accelerate the registration process for 'safer,' reduced risk pesticides; and establish separate programs for minor uses and antimicrobials. Minor uses of pesticides are generally defined as uses for which pesticide product sales are low enough to make it difficult for a manufacturer (i.e., the registrant) to justify the costs of developing and maintaining EPA registrations. Collectively, such minor crops are very important to a healthy diet and include many fruits and vegetables. Antimicrobial pesticides are substances used to control harmful microorganisms including bacteria, viruses or fungi on inanimate objects and surfaces. Antimicrobial products have traditionally included disinfectants, sanitizers, sterilizers, antiseptics, and germicides.

The most significant reforms brought about by FQPA (in terms of human health risk assessment) are on the FFDCA side and revolve around the broad changes to the tolerance setting procedures. FQPA establishes a single, health-based safety standard for pesticide tolerances and provides special considerations for infants and children. Prior to FQPA, EPA operated under a dual safety standard where raw and some processed foods were regulated under different sections of FFDCA, each with its own standard of safety; this was the old Delaney Clause. The FQPA provisions revolving around the special considerations for children are dramatically impacting the way the pesticide program assesses risk. Specifically, FQPA directs the Agency to use an extra 10-fold safety factor to take into account potential pre- and postnatal developmental toxicity and completeness of the data with respect to exposure and toxicity to infants and children; a different safety factor may be used only if, on the basis of reliable data, such a factor will be safe for infants and children. Additionally, the new law directs EPA to consider available information on: aggregate exposure from all nonoccupational sources (i.e., oral exposure from food and drinking water and dermal and inhalation exposure from pesticides used in and around the home); the effects of cumulative exposure to the pesticide and other substances with common mechanisms of toxicity; the effects of *in utero* exposure; and the potential for endocrine disrupting effects.

Incorporating these factors into the tolerance setting process poses a challenge, largely because the risk assessment methodologies and science policies for handling the additional 10-fold safety factor, aggregating exposure, and looking at common mechanisms of toxicity were not in place the day FQPA was signed, and the new law did not provide a phase-in period.

Meeting the Challenge

When FQPA was passed, EPA took an aggressive approach to implementation, with a strong emphasis on public and stakeholder involvement. The following standing committees, which all operate under the rules of the Federal Advisory Committee Act *(5)*, are providing advice to EPA on FQPA matters: (1) The FIFRA Scientific Advisory Panel and the Agency's Science Advisory Board, external peer review groups made up of experts in key scientific and public health disciplines, advise EPA on major scientific issues; (2) The Pesticide Program Dialogue Committee, a broadly representative committee, provides advice and guidance to the Agency on regulatory development and reform initiatives as well as public policy and regulatory issues associated with evaluating and reducing risks from pesticide use; and (3) The Endocrine Disruptors Screening and Testing Advisory Committee provides advice and counsel to the Agency on a strategy to screen and test endocrine disrupting chemicals and pesticides.

In addition to this very direct public involvement, EPA has been consulting with other government agencies including the U.S. Department of Agriculture, the U.S. Department of Health and Human Services, and the U.S. Department of Justice on a range of issues relevant to improving food safety, as directed by the new law.

Addressing the Scientific Provisions

As stated previously, implementing the scientific provisions of FQPA presents quite a challenge to the Agency. In broad terms, the mandates of this new law require us to dramatically strengthen the way in which we conduct our risk assessments. As defined by the National Research Council, risk assessment can be thought of as four steps *(6)*:

hazard identification; dose-response assessment; exposure assessment; and risk characterization.

During the hazard identification stage, all available toxicology data are reviewed and the endpoints (i.e., what effects the pesticide will cause) are identified. Some common endpoints include carcinogenicity, mutagenicity, and general systemic toxicity. In dose-response assessment, a reference dose or RfD is determined for toxicological effects that are believed to occur via a 'threshold' model (i.e., the effect occurs once a certain dose is met). A reference dose is an estimate of the level of daily exposure to a pesticide residue, which, over a 70-year life span, is believed to have no significant deleterious effects. Other federal agencies, such as the Food and Drug Administration, refer to this as an acceptable daily intake or ADI. The pesticide program calculates a reference dose by dividing the no-observed-effect level from a chronic study by two uncertainty factors – a 10-fold factor to account for uncertainty in extrapolating from animals to humans (i.e., interspecies) and a 10-fold factor to account for the variation within the human population (i.e., intraspecies). Exposure assessment involves determining how much pesticide humans are exposed to by the three routes: oral, dermal, and inhalation. Finally, risk characterization is the process of combining the dose-response and exposure information to describe the overall magnitude of the public health impact. Risk characterization can be expressed quantitatively via a margin-of-exposure or MOE; it includes a discussion of the uncertainties inherent to the hazard and exposure assessments.

The Additional Safety Factor. One of the more significant and interesting aspects of the new law is the additional 10-fold safety factor or 'FQPA Factor' provision. FQPA states: "In the case of threshold effects...an additional tenfold margin of safety for the pesticide chemical residue and other sources of exposure shall be applied for infants and children to take into account potential pre- and post-natal toxicity and completeness of the data with respect to exposure and toxicity to infants and children. Notwithstanding such requirement for an additional margin of safety, the Administrator may use a different margin of safety for the pesticide chemical residue only if, on the basis of reliable data, such margin will be safe for infants and children" *(7)*.

Over the past year and a half, EPA has been striving to implement this provision of FQPA. Early on, FQPA Factor decisions were made at the dose-response assessment stage and were largely driven by the completeness of the toxicological database and whether or not children appeared to be or were more likely to be more sensitive to the pesticide under question. Now, as the pesticide program has given this provision more thought and other parts of the Agency and the federal government have taken an interest in this issue, the Agency is considering moving the FQPA decision point from dose-response to risk characterization. In the same discussions, EPA is also trying to establish criteria for more widely applying the FQPA Factor – that is, applying the factor based on the entire toxicity profile and exposure scenarios, using a weight-of-the-evidence approach, and not just on the basis of increased susceptibility.

Aggregate Exposure. Another significant provision of FQPA is the requirement for 'aggregate' exposure. Under the law, "In establishing...a tolerance...for a pesticide chemical residue, the Administrator shall consider...available information concerning the aggregate exposure levels of consumers (and major identifiable subgroups of consumers) to the pesticide chemical residue and to other related substances, including dietary exposure under the tolerance and all other tolerances in effect for the pesticide chemical residue, and exposure from other non-occupational sources" *(8)*. The Agency has interpreted this provision to mean that exposures for tolerance assessments must be aggregated across all routes – oral (from food and drinking water); dermal (from residential exposure); and inhalation (from residential exposure). Prior to FQPA, in assessing exposures for tolerances, the pesticide program only considered the oral route, and generally only food, but not drinking water.

The pesticide program is faced with two challenges in implementing the aggregate exposure provision: (1) determining pesticide residues in drinking water; and (2) adequately estimating pesticide residues that occur as a result of residential exposure. Drinking water residues are difficult to discern because the Agency typically does not require these types of data. Although available from various sources (U.S. Geological Survey, states, and academia) monitoring data for pesticide residues in ground and surface water are extremely limited and often provide information on ambient water quality only and not drinking water quality, *per se*. Because of this lack of actual pesticide monitoring data for drinking water and the limited nature of ambient water quality data for pesticides, the Agency currently uses simulation models, which utilize conservative assumptions (health-protective), as screening tools to estimate pesticide residues in ground and surface water. To address the problem of estimating pesticide residues in drinking water, EPA has been working with the International Life Sciences Institute, a nonprofit, worldwide foundation established in 1978 to advance the understanding of scientific issues related to nutrition, food safety, toxicology, and the environment. A Workshop has been organized that will include participation from scientists with expertise in fate, transport, and occurrence of pesticides in ground and surface water. The working group will be asked to identify and critique currently available methods to estimate concentrations of pesticides in ground and surface waters using models and monitoring data.

Calculating reasonable residential exposure is problematic because again, these data are not routinely required and the pesticide program's current methods for estimating residential exposure tend to greatly overestimate the actual exposure. To secure better residential exposure data, the Office of Pesticide Programs is working with industry in the development of indoor and outdoor residential exposure data. Currently, the Outdoor Residential Exposure Task Force is developing exposure data for professional and nonprofessionals (i.e., homeowners) who use pesticides on turf or lawns and for exposure incurred while coming into contact with a lawn that has been treated. Additionally, industry is now considering a similar effort to generate indoor residential exposure data.

Common Mechanism. A third important scientific consideration stemming from FQPA is the 'common mechanism' provision. According to FQPA, "In establishing...a tolerance...for a pesticide chemical residue, the Administrator shall consider...available information concerning the cumulative effects of such residues and other substances that have a common mechanism of toxicity" *(9)*. The classic group of pesticides that are believed to act via a common mechanism of toxicity are the organophosphates. As with the drinking water issue, EPA has enlisted the assistance of the International Life Sciences Institute to define what is a common mechanism and then to figure out how to 'add up' the cumulative effects.

Endocrine Disruptors. The final major FQPA scientific provision relates to endocrine disruptors, where the Agency is required to develop a screening and testing program to determine whether certain substances (including all pesticides) "may have an effect in humans that is similar to an effect produced by a naturally occurring estrogen, or such other endocrine effect..." *(10)*. So far, EPA has: (1) Formed the Endocrine Disruptors Screening and Testing Advisory Committee to provide recommendations on a strategy for selecting and setting priorities for screening and testing; on identifying new/existing screening tests for validation; on the use of available screens; and on how to determine the need to test beyond the initial screens; (2) Established a national research strategy and a $10 million research program through the Agency's Office of Research and Development; and (3) Established international cooperation through the Organization for Economic and Cooperative Development and other international organizations to increase research and develop harmonized approaches to endocrine disruption issues.

Conclusion

The advent of FQPA has presented the Agency with some challenging and fascinating issues to tackle. In response to this challenge, EPA has enlisted the talents of industry, academia, environmental groups, etc., as well as our own professional staff and staff of other government agencies, to devise new and creative methodologies for fully implementing the provisions of this new law. In the years to come, we hope to continually guarantee the children of the United States a safe and healthful food supply.

Literature Cited

(1) *Food Quality Protection Act of 1996,* Public Law 104-170, August 3, 1996.
(2) *Federal Food, Drug, and Cosmetic Act*, 21 U.S.C. §§ 306b(d)(1)(H), 348(c)(3)(A), 376(b)(5)(B), 1994.
(3) National Research Council. *Pesticides in the Diets of Infants and Children*; National Academy: Washington, DC, 1993; p 359.
(4) National Research Council. *Pesticides in the Diets of Infants and Children*; National Academy: Washington, DC, 1993; p 9.
(5) *Federal Advisory Committee Act*, 5 U.S.C. App. 2.
(6) National Research Council. *Risk Assessment in the Federal Government: Managing the Process*; National Academy: Washington, DC, 1983; p 38.
(7) *Federal Food, Drug, and Cosmetic Act*, § 408(b)(2)(C)(ii), 1996.
(8) *Federal Food, Drug, and Cosmetic Act*, § 408(b)(2)(D)(vi), 1996.
(9) *Federal Food, Drug, and Cosmetic Act*, § 408(b)(2)(D)(v), 1996.
(10) *Federal Food, Drug, and Cosmetic Act*, § 408(p)(1), 1996.

Chapter 14

Impact of the Food Quality Protection Act on Industry: An Illustration Using Case Studies

C. Barrow[1], B. Shurdut[2], and D. Eisenbrandt[2]

[1]Dow AgroSciences, 1776 Eye Street, N.W., Washington, CD 20006
[2]Dow AgroSciences, 9330 Zionsville Road, Indianapolis, IN 46268

The Food Quality Protection Act (FQPA) of 1996 resulted in a number of changes that impact the regulation of pesticides in the United States. Among the most significant of these is the establishment of a single, health based standard for all pesticide residues in food. Residue concentrations in food must be determined to be "safe" which is defined as "a reasonable certainty that no harm will result from aggregate exposure to the pesticide chemical residue, including all anticipated dietary exposures and all other exposures for which there is reliable information." This fundamental safety reform will necessitate risk assessments which are more data intensive and complicated than in the past. Obtaining more accurate exposure information, including food consumption data and pesticide use information, will become critically important as companies strive to obtain, maintain, and defend their product registrations. The wide range of critical issues that require attention since the passage of the FQPA will necessitate the strengthening of existing alliances and the formation of new ones between all stakeholders engaged in the process of regulating pesticides.

Background

The most significant and far reaching reforms of the Food Quality Protection Act (FQPA) of 1996 are described in the provisions of the Federal Food Drug and Cosmetic Act (FFDCA). Under the FFDCA, the Environmental Protection Agency (EPA) must establish tolerances, or maximum legally permissible concentrations, for pesticide residues in food.

Prior to the passage of the FQPA, the FFDCA provisions required the EPA to"protect the public health" and give necessary attention "to the necessity for the production of an adequate, wholesome, and economical food supply". For pesticides declared to pose a risk for causing cancer, the EPA used a negligible risk standard. However, this was complicated by the necessity to take into account the Delaney clause of the FFDCA. Stated simply, if a pesticide causes cancer in man or laboratory animals and is concentrated in ready-to-eat processed food at a level greater than the tolerance for the raw agricultural commodity, the Delaney clause of the FFDCA prohibited the setting of a tolerance. In the case of pesticides determined to have only non-carcinogenic effects, the EPA used a safety factor approach to ensure a safe food supply for consumers. This inconsistent and confusing method for establishing tolerances had been widely criticized, even by the EPA.

The FQPA of 1996 repealed the Delaney clause and essentially overhauled the procedures for setting tolerances. The new law established a single health based standard for setting tolerance levels for all pesticide residues. In setting a tolerance, the EPA can only do so if it is safe. "Safe" has been defined as "a reasonable certainty that no harm will result from aggregate exposure to the pesticide chemical residue, including all anticipated dietary exposures and all other exposures for which there is reliable information". There are now no differences between the manner whereby tolerances are set for raw and processed foods. The tolerance setting standards also apply equally to all risks, both cancer and non-cancer.

The new law also establishes a precedent by focusing on the safety of infants and children. Since publication of the 1993 National Academy of Sciences (NAS) report on "Pesticides in the Diets of Infants and Children", the EPA has been busy implementing various recommendations set forth in the report although the previous statutes had no mandate to do so. The FQPA explicitly requires that the EPA address risks from pesticides to infants and children and, in doing so, publish a specific determination of safety for those sub-populations prior to establishing a tolerance. When establishing a tolerance now or leaving one in effect, nine factors must be considered, as follows:

- The validity, completeness, and reliability of the available data;
- The nature of any toxic effect shown to be caused by the pesticide or its residues;
- The relationship of these results to human risk;
- The dietary consumption patterns of consumers and subgroups;
- The cumulative effect of residues and other substances with a common mechanism; of toxicity;
- The aggregate exposure of consumers to the product and other related substances (dietary and other non-occupational sources, including drinking water);
- The variability of sensitivity among major identifiable subgroups (e.g., infants and children);
- Whether the pesticide may act as an endocrine disrupter; and
- Whether the product meets scientifically recognized appropriate safety factors (generally determined through animal studies).

Consideration of these factors as part of the tolerance setting process is very problematical. Data, information, and methodology are unavailable for many pesticides. As a result, risk assessments will become more data intensive and complicated.

Elements of the New Risk Assessment Paradigm

The risk assessment process is essential to any tolerance setting procedure. Historically, this involves the following considerations:

- Hazard (toxicology) information;
- Use pattern information;
- Evaluation of dietary exposure and risk assessment; and
- Evaluation of non-dietary exposure and risk assessment.

Prior to the passage of the FQPA these steps proceeded independently. The EPA estimated total dietary exposure from all foods having pesticide residues in question. Even though additional exposure pathways (i.e., drinking water or residential use) were independently evaluated, the overall risk was not combined or aggregated. The FQPA now mandates evaluation of dietary exposures (food plus drinking water) and other non-occupational exposures (residential, lawn, and garden) into a total aggregated risk assessment for a given pesticide.

In addition, if two or more pesticides share the same mechanism of toxicity and there is concurrent exposure, an increased risk to human health may occur through the cumulative effects of the pesticides in question. These cumulative effects must now be routinely factored into the overall risk assessment and tolerance setting process. These two new requirements (aggregate exposure and common mechanism of toxicity) represent significant, perhaps even monumental changes to the regulation of pesticides, both for registrants and the EPA. However, there are significant opportunities for innovative, science-based approaches. The following case studies exemplify how registrants have addressed common mechanisms of toxicity and aggregate exposure for implementation of the FQPA.

Addressing Aggregate Exposure; Chlorpyrifos as an Example

Chlorpyrifos is an organophosphorus insecticide present in many pest control products used in and around the home. Additionally, chlorpyrifos is used on many agricultural commodities. By incorporating actual exposure data, information on exposure factors and exposure groups, and market use data, a calendar-based integrated exposure model was developed which provides a realistic estimate of population-based exposures (*1, 2*). This model may be applied to any chemical (for example, a pesticide or a solvent) where there is a probability that exposures from more than one use or more than one route of exposure could occur simultaneously.

The integrated exposure assessment consisted of two basic steps:

1. Estimating exposures attributed to use of individual products; and
2. Combining these exposure estimates with the probability that more than one exposure may occur at the same time.

Inhalation, dermal and oral (for children only) routes of exposure were assessed during and following termite, lawn, and crack and crevice treatments. These exposures take into account the primary use patterns for chlorpyrifos containing products. A constant background dietary exposure was added to the model assuming that the range of daily dietary exposures can be represented by a consistent distribution of individual food consumption patterns and residues in food. Due to the small number of chlorpyrifos detections found in water monitoring studies, this route was concluded to be negligible and excluded from the assessment.

Conservative estimates of exposure were used in this assessment (*2*). For example, it was assumed that a resident would apply chlorpyrifos containing crack and crevice products 6 times a year when, in fact, market use data suggest that a resident would apply these products 2 to 3 times a year. It also was assumed that chlorpyrifos residues on floor surfaces following an application to crack and crevices would be similar to that following broadcast application. Estimates were made of integrated exposure for all children ages 0 to 6 instead of for both the < 1 year and 1-6 years age categories. Combining these two groups overestimates exposure because of the lower body weight and less surface contact time of infants less than 1 year of age. For lawn treatment, it was assumed that a liquid formulation product was applied when, in fact, it is known that as much as 50% of lawn treatments are applied as a granular formulation. Granular formulations are expected to result in significantly lower exposures, especially during application of the product and reentry on the treated turf.

In spite of these worst-case assumptions, the model provided a more realistic approach to exposure assessment than simply adding high-end exposures from all potential uses. Monte Carlo techniques were used to sample from each exposure distribution. This method, while summing exposures for all potential uses, randomly samples from the individual exposure distributions. The likelihood of having more than one exposure at any given time was determined by adjusting the distributions according to market size and frequency of use.

Exposures resulting from all uses and routes of exposure were estimated for each day of the year. These individual daily distributions were sampled to obtain an overall distribution of daily exposures that is appropriate for an entire year. Therefore, the distribution includes days when the probability of multiple exposures is likely to be higher (e.g., in the spring when lawn treatments are likely to occur) and days when the probability of exposure from multiple uses is lower (e.g., in the winter). Since most exposures are of short duration and will occur infrequently throughout the course of the year, the annual distribution of daily exposures was compared to the acute toxicologic endpoint for chlorpyrifos.

This assessment included only chlorpyrifos users who:

1. Eat food that could contain chlorpyrifos residues;
2. Have their residence treated for termites;
3. Treat their lawns and reenter treated lawns; and
4. Apply crack and crevice sprays with chlorpyrifos-containing products and reenter treated areas.

That is, a "user" was assumed to be a resident in a location where all four of these activities occur within the year. In addition, based on market use information, this "user" population of adults was estimated to include those who treated for indoor and outdoor pests themselves or by professional applicators. For self-application, it was assumed that adults were exposed during application and during subsequent reentry onto the treated surfaces.

Figure 1 represents hypothetical temporal distributions of exposures to several chlorpyrifos containing products for persons who have their residences treated for termites, consume food with chlorpyrifos residues, and have their lawns and homes treated in the same year (*2*). The fraction of the U.S. population with all of these characteristics is small and can be determined more precisely with market use data. Exposures were first calculated for "users" and then based upon market use data, extrapolated to the entire U.S. population.

A calendar based probability model provided a realistic assessment of exposures because it is unrealistic that a person will be exposed assuming maximum values for all exposure measurements and factors (*3*). Furthermore, it is very unlikely that a person will experience maximum exposures to more than one product or use simultaneously. The use of Monte Carlo techniques allows random sampling from various single-use exposure distributions and then summation of these values to form an integrated exposure distribution (Figure 2). For each exposure scenario, a single exposure value was drawn from each conditional exposure distribution based on the likelihood of occurring on a given day. A conditional exposure distribution combines exposure values with the market based probability that exposures from a specific treatment or use are present on that day. Using a commercially available Monte Carlo program (e.g., Crystal Ball®), exposure measurements were repeatedly sampled for each day of the year depending on the likelihood or conditional probability of occurring on a given day. The stored conditional integrated exposure values were combined into a single distribution of exposures and repeated for every day of the year to generate a distribution of daily exposures for "users" over the course of a year.

The input parameters necessary to run the model for a single treatment or use generally fall into four categories:

- personal exposure measurements/environmental concentration data on days before, during and after treatment.
- exposure factors such as body weight and breathing rate.
- personal and environmental factors such as activity patterns, residence type and probability that a person applies the product himself or has a professional apply the product.
- market use data that provide information on market share of chlorpyrifos containing products, frequency of application and amount of product applied.

Distributions of input parameters were used in the model where sufficient data were available, otherwise point estimates were used.

Exposure estimates were calculated for two groups, adults (both male and female) and children ages 0 to 6 (both male and female). The exposure distributions for these

Figure 1. Hypothetical Time Distribution of Exposure to 4 Use Pattern Scenarios

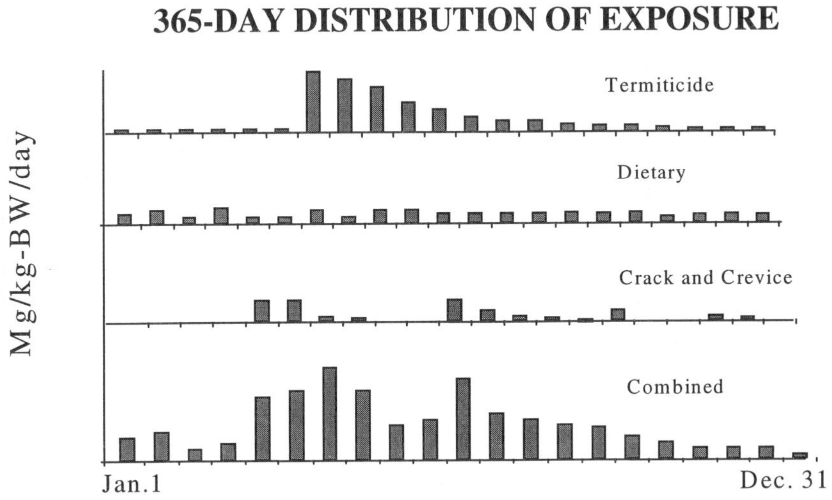

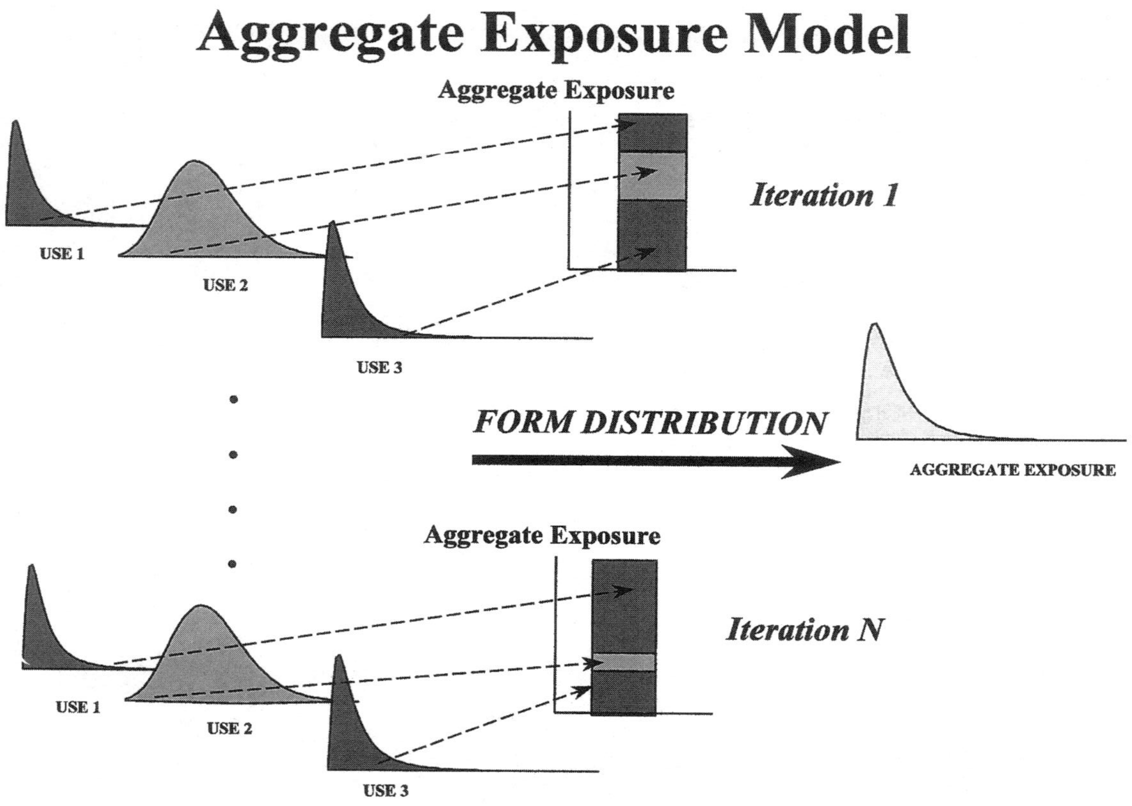
Aggregate Exposure Model
Aggregate Exposure
USE 1
USE 2
USE 3
Iteration 1
FORM DISTRIBUTION
AGGREGATE EXPOSURE
Aggregate Exposure
USE 1
USE 2
USE 3
Iteration N

"users" were compared to the chlorpyrifos acute No Observed Effect Level (NOEL) for plasma cholinesterase depression (0.1 mg/kg BW/day) or for red blood cell (RBC) depression (0.5 mg/kg BW/day) as depicted in Figure 3.

The median exposure value for adults was 9.2×10^{-5} mg/kg BW/day, and the 95th percentile is 4.6×10^{-4} mg/kg BW/day. This exposure distribution represents at most 8.8% of the U.S. adult population. The median exposure value for children was 2.4×10^{-4} mg/kg BW/day, and the 95th percentile was 1.2×10^{-3} mg/kg BW/day. Similar to adults, this exposure distribution for children represents at most 8.8% of the U.S. population ages 0 to 6. The percentages were derived assuming that for all households the residents include two adults and one child age 0 to 6 years.

The results of this assessment demonstrate that exposures likely to result from termite, crack and crevice and lawn treatments, in addition to those from consumption of foods which may contain chlorpyrifos residues are unlikely to exceed 1.2×10^{-3} mg/kg BW/day for more than 0.44% of the entire U.S. population of children and are unlikely to exceed 4.6×10^{-4} mg/kg BW/day for more than 0.44% of the entire U.S. population of adults (*2*).

This model takes advantage of the wealth of exposure and market use information available for chlorpyrifos. The result is a more realistic exposure assessment which allows an estimate of aggregate exposure based upon the likelihood of being exposed to chlorpyrifos from concurrent uses. In addition, a range of potential exposures for several populations of interest may be determined utilizing this methodology which may not be possible by simply adding high end exposures. Furthermore, since a distributional analysis generates potential exposures for a given population, an acceptable risk may be determined by using a toxicological endpoint as a cutoff at a given percentile rather than the traditional Margin of Exposure (MOE) approach.

Common Mechanism of Toxicity – Triclopyr

Criteria and guiding principles remain to be developed concerning how registrants must address the question of whether two chemicals exert toxicity through a common mechanism of toxicity. In the meantime, EPA has proposed the use of a weight-of-evidence approach to determine whether two or more pesticide chemicals are acting by a common mechanism of toxicity (*4*). Included in the guidance document as the components of hazard assessment are (1) structure-activity relationships (SAR), (2) toxicity testing, and (3) metabolism, mechanistic studies and any other chemical and biological information bearing on the mechanism of action.

An evaluation of whether a common mechanism of toxicity should be taken into account can be illustrated with triclopyr (3,5,6-trichloro-2-pyridinyloxyacetic acid) which is a pyridinyloxyacetic acid herbicide (*5*). The only other herbicide in this class of chemistry is fluroxypyr (4-amino-3,5-dichloro-6-fluoro-pyridinyloxyacetic acid). Two additional herbicides, clopyralid (3,6-dichloro-2-pyridinecarboxylic acid) and picloram (4-amino-3,5,6-trichloro-2-pyridinecarboxylic acid) from the pridinecarboxylic acid class were also included in the evaluation.

Although, the specific biochemical mechanisms of toxicity are not known, the mammalian toxicity of the pyridinyloxyacetic acids (triclopyr and fluroxypyr) and the

Figure 3. Integrated Exposure Estimate of Chlorpyrifos for Adults (Male and Female) and Children Ages 0 to 6 (Male and Female)

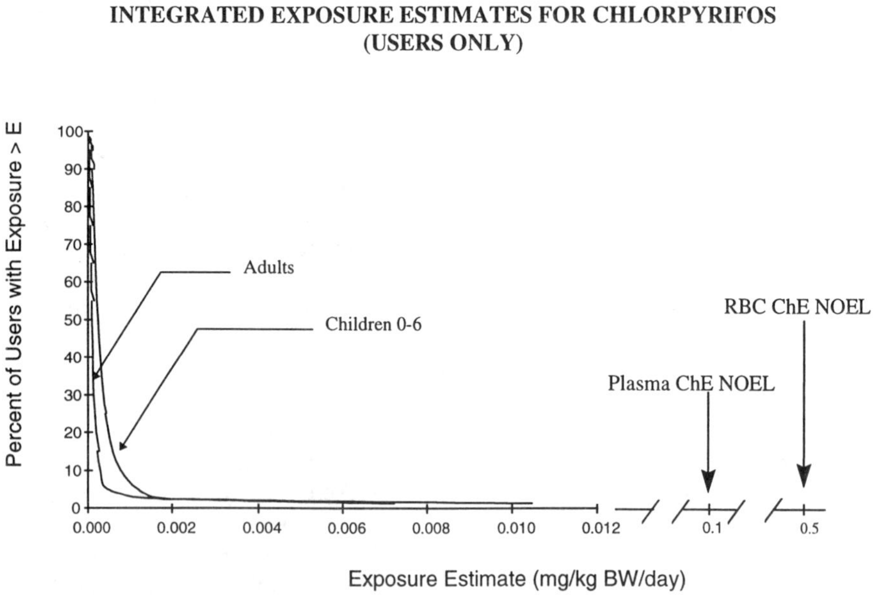

pyridinecarboxylic acids (clopyralid and picloram) have been thoroughly studied as the result of the registration and on-going reregistration process with the EPA.

The weight of evidence, based on the extensive mammalian toxicological data bases for the pyridinyloxyacetic acids (triclopyr and fluroxypyr) and the pyridinecarboxylic acids (clopyralid and picloram), indicates that there are no significant concerns, for these compounds in regard to genetic toxicity, carcinogenicity, developmental toxicity or reproductive toxicity (*6-9*). Reference doses (RfD) have been determined by regulatory agencies for these four herbicides based on an evaluation of target organ toxicity. Thus, a consideration of common mechanism of toxicity among these four compounds should proceed through a comparison of the most sensitive target organ effect.

The most sensitive target organ effects for the pyridinecarboxylic acids (hyperplasia of the stomach limiting ridge for clopyralid and liver hypertrophy for picloram) clearly distinguish these compounds from each other as well as the pyridinyloxyacetic acids which have the kidney as the most sensitive target organ. Furthermore, although triclopyr and fluroxypyr share a common target organ, the location and morphology of these changes in the kidney distinguish the effects of these compounds from each other. Dietary administration of triclopyr results in slight degeneration of a segment of the proximal tubule in the kidney of rats while chronic fluroxypyr administration is associated with a slight, diffuse increase in chronic progressive glomerulonephropathy in older animals. Overall, the toxicity testing data suggest that triclopyr does not have a common mechanism of toxicity with similar compounds.

Pharmacokinetic and biotransformation data for pyridinyloxyacetic acids (triclopyr and fluroxypyr) and the pyridinecarboxylic acids (clopyralid and picloram) show that these compounds are rapidly excreted in the urine and that the ability to excrete these materials is very large in relationship to possible exposure scenarios (*10-15*). Also, 3,5,6,-tricloro-2-pyridinol (TCP), the only metabolite of triclopyr, is low in mammalian toxicity.

To summarize, an assessment of the herbicide triclopyr indicates that there is no common mechanism of toxicity with other similar compounds as demonstrated by a comparison of structure-activity relationships, toxicity testing and pharmacokinetics. Additionally, the metabolite of triclopyr, TCP, has a low order of toxicity and is of no relevance to triclopyr's toxicity profile. Therefore, cumulative effects with other compounds are unlikely.

Conclusions

The FQPA presents a variety of new challenges to registrants and the EPA. Among the most significant are evaluations which require the need to address aggregate exposure and a common mechanism of toxicity. This new paradigm will tend to discourage reliance on default assumptions and instead will necessitate the gathering and development of more accurate exposure data including use and usage information. Specific data will almost always provide more acceptable risk assessments. Testimony to this was illustrated with the aggregate exposure assessment for chlorpyrifos.

Regarding the issue of a common mechanism of toxicity, registrants will need to advance sound, technical arguments to address the hypothesis that a common mechanism occurs. These approaches need to include the components of hazard assessment, especially those related to structure-activity relationships, available data from actual toxicity testing, metabolism and pharmacokinetics, and any mechanistic data. This approach was illustrated using the herbicide triclopyr; the conclusion being that cumulative effects with other compounds are unlikely.

Literature Cited

1. Shurdut, B.A. and Francis, M. A Case Study: Estimating the Aggregate Exposure from Major Uses of Chlorpyrifos. Presented at the Joint Society for Risk Analysis and the International Society for Exposure Analysis and Epidemiology Annual Conference, New Orleans, LA, December 8-12, 1996.
2. Shurdut, B.A. and Barraj, L.M. Evaluation of Pesticide Aggregate Exposures Under the Food Quality Protection Act: A Case Study Using Chlorpyrifos. Regul. Toxicol. Pharmacol (submitted).
3. Barraj, L.M. and Francis, M. Using a Calendar Model for Estimating the Probability of Exposure to a Chemical with Multiple Uses. Presented at the Joint Society for Risk Analysis and the International Society for Exposure Analysis and Epidemiology Annual Conference, New Orleans, LA, December 8-12, 1996.
4. EPA. Guidance for Establishing a Common Mechanism of Toxicity for Use in Combined Risk Assessments, February 11, 1997a..
5. Eisenbrandt, D.L., Nolan, R.J., McMaster, S.A., and McKendry, L.H. An Assessment of the Common Mechanism of Toxicity for Triclopyr Herbicide. Society of Toxicology Annual Meeting, Seattle, Washington, March, 1998. The Toxicologist 42: 157, 1998.
6. EPA. Reregistration Eligibility Decision (RED): Picloram. Prevention, Pesticides and Toxic Substances, U.S. EPA, EPA 738-R95-019, August, 1995.
7. EPA. Clopyralid; Pesticide Tolerance. Federal Register, Volume 62, Number 73, pp.18528-18532, April 16, 1997b.
8. EPA. Triclopyr; Pesticide Tolerances for Emergency Exemptions. Federal Register, Volume 62, Number 172, pp. 46888-46894, September 5, 1997c.
9. European Commission, Peer Review Programme. Full Report on Fluroxypyr. Pesticides Safety Directorate/EEC-Team, 5351/ECCO/PSD/97, April 9, 1997.
10. Bosch, A. Metabolism of ^{14}C-3,6-Dichloropicolinic Acid in Rats. Internal Report, The Toxicology Research Laboratory, The Dow Chemical Company, January 31, 1991.
11. Carmichael, N.G., Nolan, R.J., Perkins, J.M., Davies, R., and Warrington, S.J. Oral and Dermal Pharmacokinetics of Triclopyr in Human Volunteers. Human Toxicol. 8: 431-437, 1989.
12. Nolan, R.J., Freshour, N.L., Kastl, P.E., and Saunders, J.H. Pharmacokinetics of Picloram in Male Volunteers. Toxicology and Applied Pharmacology 76: 264-269, 1984.

13. Reitz, R.H., Dryzga, M.D., Helmer, D.C., and Kastl, P.E. Picloram: General Metabolism Studies in Female Fischer 344 Rats. Internal Report, The Toxicology Research Laboratory, Health and Environmental Sciences, The Dow Chemical Company, August 10, 1989.
14. Timchalk, C., Dryzga, M.D., and Kastle, P.E. Pharmacokinetics and metabolism of triclopyr (3,5,6-trichloro-2-pyridinyloxyacetic acid) in Fischer 344 rats. Toxicology, 62: 71-87, 1990.
15. Veenstra, G.E. and Hermann, E.A. Fluroxypyr Acid: Pharmacokinetic Study in Male Wistar Rats. Internal Report, Health and Environmental Sciences, Toxicology Research Laboratory, The Dow Chemical Company, June 1, 1983.

Chapter 15

The Food Quality Protection Act of 1996: An Industry Perspective

Mark W. Galley

Agricultural Products Research Division, American Cyanamid Company, P.O. Box 400, Princeton, NJ 08543–0400

There are two major laws that regulate the use of pesticides in the United States. These laws are the Federal Food, Drug and Cosmetic Act (FFDCA) and the Federal Insecticide, Fungicide and Rodenticide Act (FIFRA). On August 3, 1996, President Clinton signed into law major changes in these laws, which were made under the name "Food Quality Protection Act of 1996" or FQPA. Many new amendments were added to the FFDCA and FIFRA. The major theme of these multiple changes was to safen the food supply for all U.S. people but with special emphasis on children, especially the youngest and most vulnerable. It will not be possible to detail all the changes made to these laws in this short paper. Therefore, I will present what I believe are the most significant and challenging changes to the pesticide industry from FFDCA.

The FFDCA is for the most part administrated by the Food and Drug Administration (FDA) except for the control of pesticide residues in the crop food supply. This power was transferred to the Environmental Protection Agency (EPA) in 1970 (*1*). The Office of Pesticide Programs (OPP), a subsection of EPA, was given the responsibility for the oversight of pesticides.

The FQPA significantly added to the oversight of OPP by requiring EPA to determine not only the residues in food items but also the exposure to people from the drinking water as well as exposure from dermal and inhalation exposure in and around the home and residential environs.

To add to the challenge of OPP, Congress stipulated in the new Act that the new amendments go into effect on the day that President Clinton signed the law. FDA and EPA had never before established tolerances for pesticides that have to consider the aggregate exposure of pesticides in drinking water, or for dermal or inhalation exposure. Thus, they had no regulations, data requirements or protocols prepared to assist the regulated community in providing the necessary information. This timing requirement has left both the OPP and the pesticide industry struggling to comply with the new law over the last 15 months.

Under the FQPA, Congress requires EPA to review all the existing tolerances and make sure the levels of pesticide residues are safe according to the new safety standard of "a reasonable certainty that no harm will result from aggregate exposure"; which includes evaluating each of the approved and registered active ingredients (approximately 470 in agriculture) and all the crop tolerances that had been previously approved as safe under previous law within a ten-year period or by 2006.

Since there are over 9,000 of these tolerances, the Congress required that OPP carry out the tolerance reassessment in stages over a ten-year period (8/3/96-8/3/06). The Congress directed the OPP to evaluate a third of these tolerances within the first three years following the signing of the law, and further, OPP was required to choose those pesticides that they determined were likely to have the most toxic effects on the most sensitive subpopulation, that is, young children.

The OPP has chosen which group of pesticides it will examine first. The classes of chemical active ingredients known as organophosphates, carbamates, and all types of active ingredients known as B-2 oncogens (the most likely potential cancer causing chemicals). There are 39 organophosphates. The OPP has started to evaluate the approximately 1,400 established tolerances for this class of chemistry. The aggregate review of this class is scheduled by OPP to be complete by December 1998.

The new law does not allow the OPP to use the massive benefits to the production of crops in the United States to balance any potential risks from the very low levels of exposure from all sources to the general population (except in very limited cases) or its presumed most sensitive subpopulation of young children. Thus, the way OPP will decide whether all the existing tolerances and other new exposures are safe is by the following process.

The first step is for OPP to review all the toxicity information on a given organophosphate active ingredient. Fortunately, the OPP has, for the most part, a recent animal toxicity database available for each active ingredient from the pesticide industry supplying the data under the 1988 amendments to these laws. From these data, OPP chooses the lowest level of the active ingredient that can be determined to cause no harm to any of the animals tested (dogs, mice, rats, etc.) or the No Observable Adverse Effect Level (NOAEL or NOEL).

The OPP lowers the NOEL by an uncertainty factor (UF) of 10 to cover the possibility that humans might be more susceptible to the effect than the most sensitive animal species tested. The OPP again lowers the already adjusted NOEL by an additional 10 factor to ensure that in a case where, for some unknown reason, a subpopulation of humans is more susceptible to the effect it will be protected (*2*). Mathematically this can be represented as NOEL/100. This amount of the pesticide usually expressed in microgram to milligram levels available in the human diet on a daily basis (FDA refers to this as the Allowable Daily Intake, ADI) is often thought of as a volume and pictured as a measuring cup (Figure 1).

If the estimated exposure to the active ingredient or pesticide is equal to or less than the adjusted NOEL, known as the reference dose, **EXPOSURE < or = TOXICITY RfD = NOEL / U.F.'s** then there is "a reasonable certainty that no harm will result from aggregate exposure" from the approved uses, crop as well as noncrop of the active ingredient. This equation is the classic representation of risk.

The new amendments to the law instruct the OPP to add an additional uncertainty factor of up to 10 for young children, who might be more sensitive to the active ingredient, if there is any reliable information available that indicates that the young may be more sensitive than the general human population to the effects of the chemical. The pesticide industry provides data on teratology, reproduction and neurotoxicity as well as chronic studies on rodents and non-rodents that is helpful to the OPP in making the decision whether to use all or part of this uncertainty factor. So far over the last 15 months the Health Effects Division has added at least some of this additional factor if there was insufficient or additional information required to complete its aggregate risk determination .

Recently, the OPP summarized its decisions in 44 cases that have been decided since FQPA was approved (*3*). Based on thirty-two tolerance decisions, no additional factors beyond the 100 already applied to the NOEL were required. However, of the remaining 12 cases an additional 10 factor was needed in seven cases, an additional 3 factor was used in four cases, and in the remaining case a 2 factor was used. When these additional factors were added it brought the NOEL adjustment up to a thousandfold safening over the originally chosen NOEL, which in many cases was at least an order of magnitude below the Lowest Observable Effect Level or LOEL (*4*).

Prior to FQPA, once the NOEL from a given amount of a pesticide was determined (this was almost always from review of a potential chronic effect on representative adult animals), the amount of the exposure from the residues in foods with established and approved tolerances (tolerance is the maximum residue level of a pesticide that legally can be present in or on raw agricultural commodities, food or feed transported in interstate commerce) was compared to the reference dose or risk cup, as it has become popular to imagine. If, for example, the exposure in mg per kg human body weight per day is less than or equal to the allowable reference dose, or allowable daily intake, the aggregate risk assessment would allow the continued approved tolerances and labeled uses.

However, FQPA requires the OPP to now calculate the additional exposures of the given active ingredient from drinking water and from the dermal and inhalation routes (Figure 2). Since FQPA excludes the professional pesticide handlers as these people are covered by FIFRA, the exposure is calculated from noncrop uses that are present to people in and around their immediate environments; such as, home, garden, lawns, and other locations where exposure is possible due to approved labeled uses of the pesticide.

Considerable time has been given by OPP and the regulated industry, especially those registrants of the active ingredients that have been chosen to be the first to be reevaluated under the new law, to determine how to quantify the amount of the pesticides from these new required routes of exposure and, once calculated and added to the food exposure, to determine if the newly created risk cups are now full or perhaps overflowing.

If the cup is less than or equal to full, then the OPP aggregate risk reevaluation is complete and no further action is required. However, if the amount of the given active ingredient exceeds the daily allowable intake, then OPP may have to take action to lower the exposure. This action may await the generation, submission and review of new toxicity or exposure studies as was the case under the 1988 amendments to FIFRA. Under those amendments, the pesticide industry was required to replace most of the older toxicology and food dietary information prior to the OPP taking action on the removal of tolerances and other labeled uses not requiring tolerances. Alternatively, the OPP may decide to take action now to immediately remove uses (*4*).

Since the additional exposures have been calculated from what the OPP calls "reasonably worst-case" information for the water, dermal and inhalation exposures, it is the hope of the pesticide industry that the OPP will allow time for the development of the type of information that can be classified as "best-case" information derived from studies required by and agreed to by both the OPP and industry before removing previously approved labeled uses from pesticide active ingredients.

The removal, precipitously, of multiple uses from a class of insect controlling pesticides like organophosphates would, in the opinion of farmers and public health officials, cause a potential disaster without available replacement technologies.

To add to the complexity of this risk assessment, the OPP, over the last several years, has been making judgments concerning the acute effects of certain pesticides and determining if a short-term exposure of a few days to a given subpopulation's food supply (most often young children) might be more important to the risk picture and, therefore, to the calculation of the tolerance for a given food commodity than a chronic exposure.

The acute risk will be emphasized more than the chronic risk for very acutely toxic chemicals like organophosphates, especially if EPA uses food consumption at the 99.9% level from the USDA diet survey for all foods legally labeled for use for a given pesticide.

For acute risk, EPA is either:

1. doing a simple assessment using the 99.9th percentile consumption for each food weight average residue data, or

2. if possible/available using the 99.9th percentile result of a Monte-Carlo analysis, which takes into account the distributions of residue and dietary consumption data.

In either method, EPA is biasing the result towards the extreme exposures of a very few individuals whose exposure may not even have been accurately reported in the diet survey. In the case of some foods we know the data were bad because the caloric intakes per day would be humanly impossible.

For acutely toxic pesticides, like organophosphates, the (new) reassessed tolerances will be based on acute data for (most generally) young children's dietary intakes and will replace the tolerances historically set from chronic diets for an average person consuming an average diet (note that children were always included as part of the average).

This new procedure is quite different and radically changes the way pesticides are regulated in the U.S. For example, this type of risk assessment assumes that to protect all the members of the subpopulation (e.g., children), the intake of each food consumed by that group that has an approved labeled use is at an extreme level that would include 99.9% of the cumulative subgroup. This 99.9th percentile value is very biased by a few data points at the high end that may not even be accurate. EPA, in other words, assumes that all the group consumes each of those food groups in the way only a minor few members, if any, of the subgroup might consume massive amounts of the individual foods per day!

This is only for certain "highly" acutely toxic pesticides. The use of the 99.9% caloric level is not prescribed by FQPA but is an OPP policy decision.

To complicate matters even further, the OPP is required to determine what is called the "cumulative" risk of pesticides that have a common mode or mechanism of toxicity. This requires a scientific judgment call as to what the most sensitive toxic effect is of all the pesticides, followed by a grouping together of all the pesticides with a similar "mode" of toxic action. The OPP, so far, has decided that if an individual chemical cannot be reasonably separated from others as to its mode of action it must be included with all similar acting pesticides. For instance, for organophosphates, the first class of chemicals to be evaluated, OPP has decided that all 39 have the same mode of action, because they inhibit cholinesterase. Many toxicologists question the validity of EPA's approach. This decision requires OPP to consider this class of pesticides together in its determination of the cumulative toxic effect of a given mode of action.

OPP has taken the advice of its Pesticide Programs Dialogue Committee, made up of approximately 25 representative groups of "stakeholders," and is determining the individual organophosphate active ingredients aggregate exposures and then as a second step will attempt to determine a cumulative risk cup and a cumulative exposure from all FQPA exposure sources to evaluate the total risk to children from this class of pesticides.

Complicating the matter further is the possibility that the OPP may decide that approximately 20 carbamates and perhaps other types of chemistries have the same general mode of action as the organophosphates and thus will have to have their exposures and risks added to an adjusted risk cup before a final risk from pesticide active ingredients with the same or similar toxicity mode of action can be determined (*4*) (Figure 3).

The big questions concerning the cumulative risk "bowl" (if you will allow me to call it that) are how does OPP calculate the size of the bowl and if OPP decides the bowl overflows with the approved uses of the pesticides with the "same mode" of toxicity, how does it go about reducing the uses in a fair and reasonable manner taking into account the user community?

Figure 1. Reference Dose

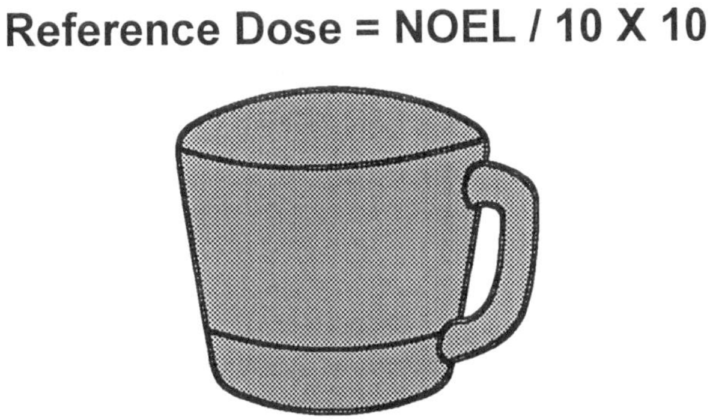

Figure 2. Post FQPA Risk Cup

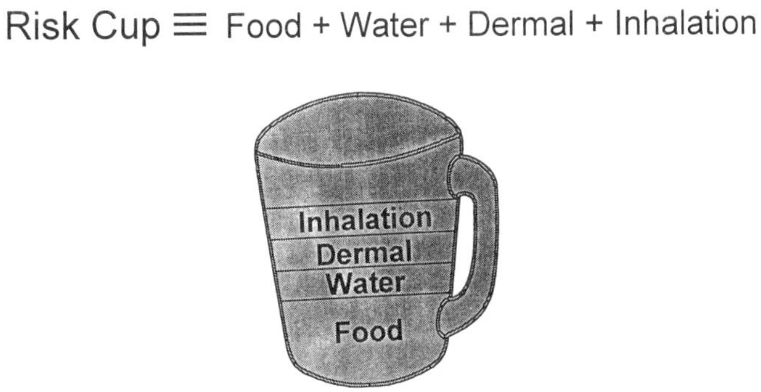

Figure 3. Cumulative Risk "Bowl"

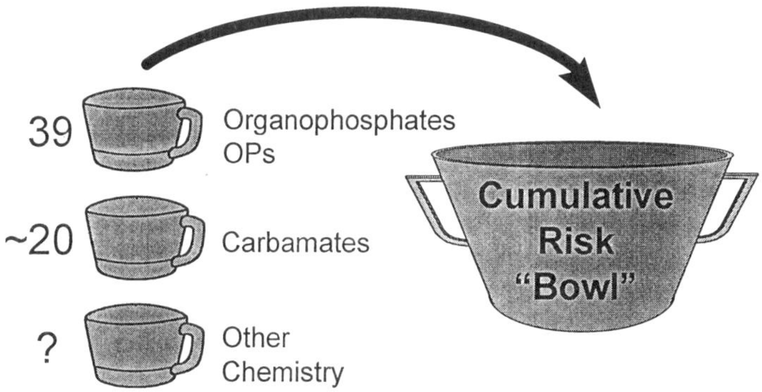

Congress decreed that the first third, worst-case tolerances had to be reassessed within 3 years. Therefore, these assessments must be carried out by OPP and the pesticide industry by August 3, 1999.

It is the hope of the pesticide industry that reason and common sense can be brought to the implementation of the FQPA. Perhaps OPP would join the pesticide industry in petitioning Congress to allow a slight time extension in the implementation of at least these few sections of the new law that I have described. Time is needed in order for the pesticide industry to be able to provide the OPP with scientifically sound data and risk assessments to make its decisions under the FQPA.

For a more complete review of the FQPA, the EPA's OPP published a Plan entitled **1996 Food Quality Protection Act: Implementation Plan** (*4*) in March 1997 and most probably will be issuing new installments, at least, on a yearly basis. All the changes to the FFDCA and FIFRA are explained in this publication. If anyone would like a copy write to the Office of Prevention, Pesticides, and Toxic Substances at 401 M Street, Washington, D.C. 20460 and they will be happy to send you a copy of their plan.

Literature Cited

1. *Pesticide Regulation Handbook*; Third Edition, McKenna & Cuneo and Technology Services Group, Inc.; 1991; Chapter 1.
2. *Pesticide Regulation Handbook*; Third Edition, McKenna & Cuneo and Technology Services Group, Inc.; 1991; Chapter 5.
3. *Presentation to the Pesticide Program Dialogue Group*; October, 1997; by the Hazard Effects Division of the Office of Pesticide Programs.
4. "*1996 Food Quality Protection Act: Implementation Plan,*" United States Environmental Protection Agency; Prevention, Pesticides and Toxic Substances; March, 1997; Part 4.

Chapter 16

Good Laboratory Practices and Pesticide Regulation in Mexico

Amada Velez

Secretaria de Agricultura, Ganaderia y Desarrollo Rural, Direccion General de Sanidad Vegetal, Direccion de Servicios y Apoyo Tecnico, Mexico, D.F., 04100

Traditionally, Mexico has adopted Codex and USEPA maximum residue limits (MRL). However, we have recently considered the need of establishing our own MRL, taking into account differences in crops, climatic conditions, agricultural practices, pests and diseases, etc. To accomplish such a responsibility, Mexican authorities have issued a regulation that establishes the procedure to conduct field trials in order to determine the maximum residue which may be expected on a raw agricultural commodity as a result of the authorized use of pesticides.

The Intersecretarial Commission for the Control of Production and Use of Pesticides, Fertilizers and Toxic Substances, integrated by four Secretaries: Agriculture, Health, Commerce and Environment, has the authority to establish maximum residue limits (MRL) in Mexico.

Mexican authorities issued a regulation (*1*) that established criteria to carry out field studies with the objective of obtaining maximum residue limits, which was introduced as follows:

Mexican Official Standard by Which the Phytosanitary Specifications and Requirements are Established for the Conduct of Field Trials for the Establishment of Maximum Pesticide Residue Limits in Agriculture Products

Goal and Scope of Application

The observance of Mexico's Official Standard is internally compulsory in the national territory. It has the purpose of establishing the requirements and specifications, as well as the proceedings and criteria upon which the field residue trials should be conducted to determine pesticide residues in raw agricultural products, in order to establish maximum pesticide residue limits. Therefore, it is applied to all pesticides on which a maximum residue limit should be established.

Any registrant who conducts a field trial for pesticide residue evaluation in agricultural products, with the purpose of establishing maximum pesticide limits, should comply with this Standard.

Specifications

Background. The Secretary, or any duly authorized person, shall inspect field trials for the establishment of maximum pesticide residue limits in compliance with Good Laboratory Practices principles.

Procedures for the Development of Field Studies Pursuing the Establishment of Maximum Pesticide Residue Limits. Registrants who conduct field trials to determine pesticide residues in plant products should give notice to the Secretary or to the authorized persons of the date of study initiation. This notice should be submitted no more than 15 days before the start of the study.

The notice for the initiation of the study should include the following information:

Study Plan. A plan should be submitted in a written form prior to the initiation of each study to the Secretary or to the authorized person. This plan should contain the following information:

- A descriptive title;
- A statement which shows the nature and purpose of the study;
- Identification of the substance by code or name (IUPAC, CAS number, etc.);
- Name and address of the sponsor;
- Name and address of the place where the test should be conducted;
- Name and address of the Study Director and/or Principal Investigator;
- Date of agreement to the study plan by signature of the Study Director or the test facility management;
- The proposed starting and completion dates;
- Test method to be used, including data and description for the experimental design; the description of. the chronological procedure for the study; all methods, material and conditions; type and frequency of the analyses; measurements, observations, and evaluations to be performed;
- Storage areas for the products to be applied and the application equipment;
- Name of the individual responsible for quality assurance;
- Laboratory for sample analysis that should be operating under the Good Laboratory Practices;
- Geographic location of the areas in which the field studies shall be conducted. Commitment letter for disposal of crops or materials generated during the performance of the study, stating the place, date and the disposal practices.

Commitment Letter. A commitment letter must br submitted for disposal of crops or materials generated during the development of the study that states the place, date and the disposal practices.

Organization and Personnel. The requirements for personnel, the project, materials, supplies, equipment and facilities should be fulfilled for the conduct of field trials for pesticide residue detection.

In pursuit of the establishment of maximum pesticide residue limits, the field trials for pesticide residue determination should be conducted under the Good Laboratory Practice Principles (GLP) that were established in the OECD's consensus document (*2*). The conditions included in that document which shall be fulfilled are as follows:

Testing Facility Management. The testing facility shall have a management unit responsible for the following:

1. Assure that the personnel, resources, facilities, materials, equipment and methodologies are available;
2. Maintain a record of qualifications, training, experience and job description for each professional and technical individual;
3. Assure that the personnel clearly understand the functions which they are to perform and, if necessary, provide technical training;
4. Assure that personnel shall take necessary personal sanitary and health precautions;
5. Assure that Standard Operating Procedures shall be established and followed;
6. Assure that there is a Quality Assurance Program with personnel assigned;
7. When appropriate, agree to the study plan in conjunction with the sponsor;
8. Assure that the study plan amendments are duly agreed to and documented;
9. Maintain copies of all study plans;
10. Maintain historical archives of all Standard Operating Procedures, providing the address at which they are located;
11. Ensure that a sufficient number of personnel is available for a timely and proper conduct of each study;
12. A scientist or other professional of appropriate education, training and experience, or combination thereof, shall be designated as the Study Director before each study is initiated. If it is necessary to replace a Study Director while the study is being performed, this should be documented; and
13. Assure that an individual is identified for the archives management.

Study Director. The Study Director shall be responsible for:

1. The overall conduct of the study and for its report;
2. Agreement for the study plan;
3. Ensuring that the procedures specified in the study plan are followed, and that authorization for any amendment is obtained and documented together with the reasons for them;

4. Ensuring that all data generated are fully documented and recorded;
5. Signing and dating the final report to indicate acceptance of responsibility for data validity and confirming the Good Laboratory Practice principles fulfillment;
6. Ensuring that once the study, the study plan, the final report, raw data and supporting material are completed, these documents are transferred to the archives;
7. Ensuring that the test and reference substances are available at the test areas when such products are necessary; and
8. Ensuring that an adequate and coordinated handling system exists between personnel assigned to the field phase of the study and the analytical laboratories for sample analysis.

When the Study Director cannot enforce his/her supervisory control during any phase of the study, a principal investigator should be designated and assigned to act in his/her place for the specific phase.

Principal Investigator. The principal investigator shall be named in the study plan or in an amendment, emphasizing the study phases under his/her responsibility. The principal investigator shall be a qualified and knowledgeable person for the supervision of the phase under his/her control.

The principal investigator shall assure that the relevant research phases are conducted according to the study plan, the related Standard Operation Procedures, and under the Good Laboratory Practice.

Quality Assurance Unit. The test facility should have a documented person in the Quality Assurance Program to assure that the studies performed are in compliance with the Good Laboratory Practices. The program should be carried out by an individual or by individuals assigned by management.

The responsibilities of the Quality Assurance Unit shall include the following:

1. Assure that the study plan and Standard Operating Procedures are available to the personnel conducting the study;
2. Assure that the study plan and Standard Operating Procedures are followed by periodic inspections of the test facility and/or by auditing the progress of the study. Records of such procedures should be retained;
3. Immediately report to the management and to the Study Director unauthorized deviations from the study plan and from Standard Operating Procedures;
4. Review the final reports to confirm that the methods, procedures and observations are accurately described, and that the reported results accurately reflect the raw data of the study; and
5. Draw up and sign a statement to be included with the final report, that specifies the dates when the inspections were carried out and the dates when any finding was reported to management and to the Study Director.

Besides these responsibilities, others could be imposed, depending on the study plan type and on the Standard Operating Procedures.

Facilities. Facilities should have the following requirements:

1. Adequate trial areas with a minimum of external interferences, free of chemical substances or where a case history of the pesticides used is available. These areas should be identified with signs or landmarks;
2. Storage areas for equipment, supplies, and test chemical substances individually separated to avoid and/or prevent any contamination. These storage areas also should be adequate to preserve the identification, concentration, purity, and stability of the test items;
3. Disposal procedures for pesticides and wastes;
4. Archive space should be provided for the storage and retrieval of raw data, reports, samples and specimens;
5. Equipment is adequately inspected, cleaned, and maintained. Equipment used for the generation, measurement, or assessment of data shall be adequately tested, calibrated, and/or standardized; and
6. Written Standard Operating Procedures approved by the Study Director should be maintained for consultation.

Study Plan Specifications for Residue Trials. A test facility should have a written plan prior to the start of the study. All amendments, changes, and reviews approved by the Study Director that are intended to ensure the quality and integrity of the data generated during the course of the study shall be documented, signed and dated by the Study Director and maintained with the original plan. The study plan should contain, but should not be limited to, the stipulated information listed previously.

Criteria for Study Conduct. The general criteria under which residue trials should be performed are:

1. Two trials should be performed in geographical areas, noting cycles, and using agricultural practices representing the crop and the region;
2. When a product is applied to a crop close to maturity, studies on decreasing residue levels are needed to determine the acceptable preharvest intervals;
3. The field trials should be conducted with proposed commercial formulations and not with formulations prepared at the laboratory;
4. The product application should be carried out with commercial equipment in analogous practices as those used by farmers or with equipment simulating common agricultural practices;
5. The trials should be carried out to determine and evaluate the conditions and factors that lead to the highest residue level after following the recommended use patterns;
6. When a chemical product is applied to a harvested crop, information should be obtained on alteration of the amounts and nature of the residue during the normal course of storage and handling of the crop after treatment;

7. Considering the large variety of crops and agricultural products on which a pesticide may be used, it may not always be necessary to carry out trials on all crops, but only on a representative commodity of a group with equal phenotypical characteristics whenever agricultural practices are the same; and
8. In cases where the product under study is applied to non-edible crops, residue data will not be necessary.

Technical Criteria Under Which the Field Trials Should be Carried Out.

Trial Lay-out. Selection of trial sites should be carried out in major areas of cultivation or production and should be located in such a way as to cover the range of relevant representative conditions, such as, climatic, seasonal, edaphic, cropping system, etc., that are likely to be met for the intended use of the pesticide.

The number of sites needed depends upon the range of conditions to be covered, the uniformity of crops, and the agricultural practices.

Results of two tests from the representative agricultural areas and from the characteristic agricultural cycles should be needed.

Replication. The variations of residue levels obtained from the same place are small compared with those found in data from different sites. Therefore, it should not be necessary to replicate treatments at individual sites.

Plots. The size of the individual plot should be large enough to apply the pesticide in an accurate and realistic manner and to provide representative crop samples.

A control plot for supplying untreated samples is necessary. The control plot should be located close enough to ensure identical growing and climatic conditions but separated to exclude any contamination from treated plots due to drift, volatilization, leaching, etc.

For pesticides of high vapor pressure, such as, fumigants, aerosols, smoke, or fogs, that are used in greenhouses or in stores, the control samples from untreated or stored crops should come from greenhouses maintained under almost the same conditions.

Application of the Pesticide.

Methods of Application. The method of application should reflect the intended recommendation which should be carried out with equipment similar to that used in local practice or shall simulate the commercial application practice that is used in the zone where the study is to be conducted. The greenhouses or storage enclosures where high vapor pressure products, such as, fumigants, aerosols, smokes or fungicides to control mildew are used must be completely treated.

Dosage Rates. In a residue trial the maximum rate proposed should be applied.

Number and Timing of Applications. The number of treatments and intervals between applications should reflect the minimum and maximum use of the pesticide product which shall be recommended.

Additional Pesticides. If the application of other pesticides is necessary, the products used should be recorded in the control sheet and should not interfere with the product under study. The trial and control plots should get the same treatment regarding the additional pesticides to be used.

Representative Field Samples. The size of the sample must be representative of the trial, and also it should be taken according to Standard Operating Procedures. While performing the sampling, the following should be taken into consideration:

1. Avoid collecting diseased or under-sized crop parts or commodities at a stage when they would not normally be harvested;
2. Sample the parts of the crop that normally are the commercial commodity;
3. Collect the samples in such a way that reasonably represents the typical harvesting practice;
4. Take care not to remove surface residues during handling, packing or processing;
5. Collect and bag the required weight of samples in the field and do not sub-sample; and
6. Samples must be collected by trained personnel.

Sampling Procedures. The samples must be collected by the Study Director's assigned staff.

Primary Samples. As far as possible, primary samples should be taken from all parts of the plot. The Study Director or Principle Investigator must record every deviation from this requirement. The primary samples should be of a similar size, and the total weight should be composed of the combination of all primary samples as in bulk. The final sample should never be of a lower weight, keeping in mind the possible requirement for a new subdivision and having to provide adequate laboratory samples.

Control Samples. Control samples should be of similar quality to the test samples. Control samples should be collected before the treatment samples so as to avoid a possible contamination during their handling.

Sample Packing and Shipment. The field sample should be placed in a clean container made of inert material and adequately protected against all possible factors of external contamination and damages that could happen during its transfer. The container must be hermetically closed in such a way that an unauthorized opening could be detected and delivered to the analytical laboratory as soon as possible. All the necessary precautions must be taken against decomposition, e.g., samples liable to deteriorate must be kept refrigerated or frozen.

Labels and Records. Label each sample with adequate sample identification. Every laboratory sample must be adequately identified and must be accompanied by a data sheet showing the nature, origin of the sample, date, and sampling location, together with all other data that could be helpful for the analysis.

Deviations from the Recommended Sampling Procedure. If, for any reason, a deviation to the recommended procedure should happen, the data sheet should show all the procedural details which had been followed and applied.

Good Laboratory Practices. All samples and documents shipped to the residue laboratory should be done according to the "Good Laboratory Practices" principles as described in the Standard Operating Procedures, respectively.

All data generated during study conduct should be recorded directly and legibly in black indelible ink by the individual entering the data. These entries should be dated and signed or initialed. Any change made in the raw data sheet should be immediately justified, dated and signed.

To assure the quality and integrity of the data generated during the course of the study, the test facility should have written Standard Operating Procedures relevant to the work being performed. The SOPs should be available for, but should not be limited to, the following categories of activities:

1. Calibration of the application equipment;
2. Calibration of measuring apparatus;
3. Weighing of test substances;
4. Measuring of test substances;
5. Application of test substances;
6. Sampling;
7. Sample packing for delivery to the laboratory;
8. Washing of the application equipment;
9. Receipt, transportation, storage and dilution of test substances;
10. Establishment of plots at the field;
11. Unit of quality assurance;
12. Use and handling of environmental monitoring equipment; and
13. Recording of raw data.

Reporting of the Study Results. A final report should be drawn up for the study in two parts.

Field Report. The field report must include, at least, the following information:

1. A descriptive title;
2. Objective and procedures stated in the approved protocol, including any change in the original protocol;
3. Identification of the test substance by code or name; CAS number; characteristics of the test substance, including purity, stability, homogeneity and composition or other adequate characteristics;

4. Researcher's name;
5. Study Director's name;
6. Name of other principal personnel contributing reports to the final report;
7. Dates on which the study was initiated and completed;
8. A quality assurance statement certifying the dates when inspections were made and the dates any findings were reported to the Study Director and to the management;
9. Description of the methods and material used;
10. Copy of the raw data;
11. Listing of the SOPs used in the study; and
12. Evaluation and discussion of the results.

Laboratory Report. The laboratory report should include the following data:

1. Analytical method used for the analysis;
2. Copy of the data obtained;
3. Record of temperature at which the sample was stored during study conduct;
4. Report of results, including calculations and analytical method used for the analysis;
5. Results expressed in residue levels detected during this study;
6. Proposal for maximum pesticide residue limits;
7. Listing of the SOPs used in the study; and
8. Evaluation and discussion of the results.

Besides the information required above, additional information could be required, depending on the results in the study final report.

Conclusions

Mexico is in the process of establishing its own maximum pesticide residue limits; therefore, we have published regulation norms that establish the criteria under which field studies shall be carried out in order to determine the residue level which remains in the crop derived from a Good Agricultural Practice.

In the formulation of this regulation, international guidelines have served as its base, such as, "OECD Principles of Good Laboratory Practice" and "Guidelines on Pesticide Residue Trials to Provide Data for the Registration of Pesticides and Establishment of Maximum Residue Limits" *(3)* of the Food and Agriculture Organization of the United Nations.

The purpose of this regulation is to provide information to the petitioner on the criteria and protocols that shall be followed for the field trial data design, performance and reporting. We have harmonized our procedure for the establishment of Maximum Pesticide Residue Limits to be congruous with the international requirements.

Literature Cited

1. SAGAR. Norma Oficial Mexicana NOM-050-FITO- 1995. Por la que se establecen los requisitos y especificaciones fitosanitarias para efectuar ensayos de campo para el establecimiento de Limites Maximos de Residuos de Plaguicidas en productos agricolas, 1996.
2. The OECD Principles of Good Laboratory Practice. Environmental Monograph No. 6. The Application of the Good Laboratory Practice Principles to Field Studies. Environment Monograph No. 50 Paris, 1992.
3. Food and Agriculture Organization of the United Nations (1990) Guidelines on Producing Residue Data from Supervised Trials, 1990.

Chapter 17

Pesticides Registration Process in Mexico

J.P. Serres[1], J. Morgado[2], and G. Salas[3]

[1]Marketing, [2]R & D Department, and [3]R & D Department—Registrations, Novartis Agro, S.A. DE C.V., Mexico City, Mexico

The registration of pesticides in Mexico is a long and complex process that involves many regulating institutions from different Ministries, such as, Agriculture, Health, Ecology and Commerce.

To commercialize pesticides in Mexico, each supplier needs to have a legal presence. The first step is to be audited by an approved technician and have available the following documentation required by the different Ministries. See Table I:

Table I. Company Permits

Ministry	*Activities To Be Registered*
1. **Agriculture** (SAGAR)	
Company Registration according to the: (NOM-033-FITO-1995) (NOM-034-FITO-1995)	• Manufacturer • Formulator • Importer • Distributor
2. **Health (SSA)**	
Federal sanitary license	
3. **Ecology (SEMARNAP)**	
License to operate as a formulator.	

To obtain the above mentioned permits, a company must meet a series of specific requirements established in different Mexican laws, norms and procedures.

Metrology and Standardization Federal Law. It is under the responsibility of the Commerce Ministry through the Standardization Department and the National Committee of Standardization to promote and coordinate the metrology and standardization activities among the different departments and entities of the Public Administration (*1*).

Health Law. It is under the responsibility of the Health Ministry through the Department of Health and Labor Environment and the National Committee of Standards & Regulation and Sanitarian Protection to regulate matters on fertilizers, pesticides and toxic substances (*2*).

Federal Law On Crop Protection. It is under the responsibility of the Agriculture Ministry through the Crop Protection Department and the National Phytosanitary Council to promote and to watch over the fulfillment of the phytosanitary resolutions as well as the development of activities and phytosanitary services (*3*).

Mexican Official Standard. NOM-034-FITO-1995. It establishes the phytosanitary requirements and specifications to manufacture, formulate, toll formulate, formulate and/or toll manufacture, and import of agricultural pesticides (*4*).

Mexican Official Standard. **NOM-033-FITO-1995**. It establishes the phytosanitary requirements and specifications to commercialize agricultural pesticides (*5*).

When this documentation is obtained, a company can continue with the registration process of every single product (active ingredient and its formulations).

In Mexico, a joint Commission, **CICOPLAFEST** - *Interministerial (Commerce, Agriculture, Ecology, and Health) Committee for the control of the process and use of pesticides, fertilizers and toxic substances*, is constituted to control the registration process and, therefore, the use of pesticides, fertilizers, and toxic substances. Table II provides the name of the ministries and their functions.

This Commission has established 7 different types of registrations, according to the class of products involved:

- Plant Protection Products
- Animal Health Products
- Pesticides for Forestry
- Turf and Ornamentals
- Non-cultivated Area Products
- Domestic Pesticide (indoor)
- Active Ingredients

Table II. CICOPLAFEST Organization

Actual Members	*Functions*
SAGAR - Agriculture Ministry	Use and recommendation of pesticides
SSA - Health Ministry	Toxicological information label of pesticides
SECOFI - Commerce Ministry	Norms publication
SEMARNAP - Ecology Ministry	Ecological information
New Participants by the end of 1998	
ST - Labor Ministry	Workers' safety on pesticide handling
SCT - Communication and Transportation Ministry	Regulation of pesticide transportation

CICOPLAFEST Objectives. The objectives of CICOPLAFEST are:

- To coordinate regulation activities and control of pesticides, fertilizers and toxic substances.

- To establish procedures for the analysis and further resolution regarding the registration and authorization related to the manufacturing, formulation, transportation, final disposal of left over, containers and packaging, of pesticides.

- To issue registrations and import authorizations in accordance to the established program.

CICOPLAFEST Organization. CICOPLAFEST is composed of the following:

- A Technical Committee: Integrated by the directors of each Government Ministry mentioned above.

- A Sub- Committee: Integrated by personnel responsible for the analysis and revision of the information and documentation submitted with each application for registration.

- A Reception and follow - up area: Informs end user on status of the process for every single application.

This process involves a lot of paper work, this is inevitable, and it is caused by the number of forms to be filled out.

Information Provided. The required technical, toxicological, ecological and safety information for product registration is submitted to CICOPLAFEST, which is the organization that controls the process of registration of pesticides in Mexico, to be registered as an industrial product. Table III summarizes the information to be provided for the review of a pesticide application.

Table III. Required Information and Length of Time for Review.

Content	*Timing*
1. Registration Dossier	
• Toxicological and environmental studies	3 to 6 months
Safety (handling of product):	
⇒ If the product is a technical material, necessary worker and community warning statements must be indicated on the label in case of explosion or leakage.	
⇒ If the product is a ready formulation, necessary product application precautions must be indicated, also in text labels	
• Country of origin documentation	1 month
2. Biological Data	18 months
• Analysis and further approval of the report of the field trials: Field trials must be performed in Mexico and have to be validated by a certified inspector who ensures that the trials have been properly conducted according to procedures established in the norms (*6*). These studies must fulfill the requirements established in the official Mexican norm NOM-032-FITO-1995 which indicates the phytosanitary specifications for the development of biological efficacy studies with pesticides and its technical resolution.	
• When the active ingredient or the formulated product is new, an experimental evaluation permit must be obtained and local field trials must be conducted in different locations for biological efficacy, using different crops, and must include several rates of application.	12 months

Table III. (Continued)

3. Registration Questionnaire:	
• Although most of the documents are accepted in the English language, this document is in the Spanish language. This is a kind of summary of the complete studies included in the dossier.	1 week
4. Label text:	2 days
• The contents of the company label text must comply with official Mexican Norm (7) and use recommendations only can be included along with efficacy field trials. NOM-045-SSA1-1993. This is the case for products that include the following uses: agricultural, forestry, animal health, and gardening.	
5. Registration application form:	
• This form contains the general information. The following documents should be attached: the sanitary license, the company registration as manufacturer and formulator, the printed label of the country of origin, the registration certificate for use in country of origin (in Spanish).	1 day

Registration Procedure

When all the documents have been submitted to the authorities, CICOPLAFEST is then responsible for the analysis and further decision regarding the Application for the registration. This process takes from 3 to 4 months to be completed. When it is finally approved, it is then submitted for signature to the heads of SEMARNAP (Ecology Ministry), SAGAR (Agriculture) and SSA (Health), this process takes other 3 months. Usually, if it is not approved, additional information is required and the whole process is then repeated.

Follow-up is essential to complete the registration of products since, once the registration package has been turned in, the waiting period varies from 17-18 months.

When the dossier information is obtained, the following steps and timelines, as delineated in Table IV, are followed to complete the registration process:

Table IV. Submission Elements and Length of Time for Review.

Activity	*Timing*
• Submission of application form and dossier information	1 day
• Review and analysis of the information submitted by the Technical Subcommittee. The Subcommittee decides if the petition is approved, denied or keeps pending.. If the status is pending, a company must complete any additional information/documentation required.	5-6 Months
• Submission of additional information. Missing information is submitted for review and analysis by the Subcommittee which will issue the final approval.	1 Month
• Registration approval	1 Month
• The certificate of registration is issued by the Ministry of Health	2 Months
• The registration document has to be signed by the Directors of Crop Protection (DGSU-SAGAR), Health (DGSA-SSA), and Ecology (INE-SEMARNAP).	2 weeks

Literature Cited

1. SECOFI. 1992. Federal Law of Metrology and Standardization. Published in the Federal Official Diary on July 1st. 1992.
2. SSA. 1994. Health Law. Published in the Federal Official Diary on February 7, 1984.
3. SAGAR. 1994. Federal Law on Crop Protection. Published in the Federal Official Diary on January on January 5, 1994.
4. SAGAR. 1996. Mexican Official Standard NOM-034-FITO-1995 Published in the Federal Official Diary on June 24, 1996.
5. SAGAR. 1996. Mexican Official Standard NOM-033-FITO-1995 Published in the Federal Official Diary on June 24, 1996.
6. SEMARNAP. 1998. Federal Law on Ecology and Environment Protection. Published in the Federal Official Diary on January 28, 1998.
7. SAGAR. 1995. Mexican Official Standard. NOM-032-FITO-1995. Published in the Federal Official Diary on December 9, 1996.
8. SSA. 1995. Mexican Official Standard NOM-045-SSA1-1993 Published in the Federal Official Diary on October 20, 1995.

Chapter 18

Environmental Assessment of Pesticides in Brazil

G. S. J. Dubois, D. A. do Carmo, M. Zerbetto, and E. R. dos Santos

Departamento de Qualidade Ambiental, DIRCOF—IBAMA, SAIN Av. L4 Norte, Ed. Sede Bloco "C", Brasilia, DF CEP: 70800–200, Brasil

The environmental assessment of pesticides has been one of the steps in the process of pesticide registration in Brazil since 1990. The environmental assessment consists of the analysis of physical and chemical properties of the product, environmental fate studies, acute and chronic toxicity tests on nontarget organisms, and various other tests. The environmental assessment method currently adopted is performed by assessing potential hazards to the environment and that assessment leads to a final environmental classification that ranges from Class I to Class IV (highly to slightly dangerous products) or to products deemed to be unacceptable. The Brazilian Institute of Environment and Renewable Natural Resources (IBAMA) Act 84/96 established government entitlement to request field monitoring studies to provide more extensive data in order to establish environmental policies aimed at minimizing any environmental impact generated by the use of pesticides.

Assessing the environmental effects of pesticides is an important part of the process of developing regulatory controls for pesticide registration. This process is necessary to assure that pesticides are used in a way that will maximize their usefulness to users and minimize environmental hazard, thus ensuring for the future the protection of the agricultural ecosystems so that they can be used for continuous and sustained food and fiber production.

Development of Brazilian Legislation on Pesticides

Until 1989 the process of licensing pesticides in Brazil was under the responsibility of the Ministry of Agriculture and the Ministry of Health and was based on Decree # 24 114 issued on April 12, 1934. So, for more than half a century the basic Brazilian

Legislation on pesticides was not altered while domestic and global realities concerning the development and use of pesticides were undergoing radical changes. The amendment of legislation and, consequently, of the national registration system was obviously needed in order to adapt it to the new realities.

Decree # 24114 was replaced by Law # 7,802 of July 11, 1989 (*1*), that was regulated by Decree # 98816 of January 11, 1990 (*1*). The new Law made substantial changes in the Brazilian system of registration and control of chemical and biological products, not only for those products used in agriculture and agroforestry but also for those used in industrial sites, public spaces, household, aquatic environments, and public health campaigns. The current legislation takes a clear stand for the protection of human health and the environment and is in accordance with regulations adopted by some developed countries. It is quite comprehensive, covering not only the process of licensing but also testing guidelines (*2*), codes and rules.

Governmental actions regarding pesticides in Brazil, in addition to granting or denying permission to register products previously licensed in other countries for introduction into the Brazilian market, involve prevention of international illegal trading, quality control monitoring, control of pesticide sales by retailers, orientation and education of users and traders, processed food control monitoring, and the performance of epidemiologic and environmental monitoring. It may be safely argued that, among all the tools that integrate the process of governmental control of pesticides, registration of these chemicals is the basic step.

Pesticide Assessment Ensures a Safer Use

The pesticide licensing process gives our government authorities involved the opportunity to study and to assess the agronomic, toxicological and environmental impact of the product. The data obtained provide them with a sound scientific basis when deciding whether to grant a license or not and whether to establish additional restrictions and recommendations on the use of the product. This is necessary in order to guarantee to the public a safer use of the product and minimal hazards to man and to the environment.

In fact, the assessment of the agronomic efficacy of a product may be clearly measured in the field by its users and, consequently, will define its success or failure in the market. On the other hand, some harmful effects on human health, on natural resources, and on the quality of the environment may not be perceived by users in a short period of time, thus leaving the public or the pesticide users in a passive situation vis-a-vis these products. So, the assessment of potential negative effects of the product on human health and on the environment is of vital importance in the pesticides licensing process.

It must be stressed that the process of pesticide registration established by Law # 7802 (*1*) and Decree # 98 816 (*1*) is aimed at the protection of human health and the environment. Furthermore, licensing a potentially harmful product is a decision making process and must be based on a cost-benefit analysis involving a comprehensive range of toxicological, ecotoxicological and agronomic aspects, while always bearing in mind the overall interests of the Brazilian society.

Evaluation and Classification of the Potential Hazard to the Environment

Evaluation of the agronomic efficacy is made by the Ministry of Agriculture; the toxicological evaluation of effects on human health is made by the Ministry of Health; and evaluation of the environmental toxicity is made by IBAMA. All evaluations are based on tests and information provided by the companies that want to register a pesticide, as well as on literature and data banks.

In the specific case of pesticide registration, the basic responsibility of the environmental agency, IBAMA, is the evaluation and classification of products and substances according to their potential hazards, as established in Article 5 of Decree 98 816 (*1*).

The first IBAMA Act aimed at regulating this process was Act # 349 (*1*) of March 1990, later replaced by Act # 139 (*1*) of December 1994. Pesticide product evaluation is currently ruled by Act # 84 (*3*), issued in October 1996, which replaced Act # 139 (*1*). These Acts have been gradually altered in order to improve the pesticide evaluation process by requiring more non-target organism tests and field monitoring studies.

Period of Transition from the Evaluation of Potential Hazards to Environmental Risk Assessment

The evaluation and classification of pesticides conducted by IBAMA refer to potential hazards to the environment. There is no doubt that risk assessment is the more rational and adequate procedure to be used; however, the technical procedures used to perform it are more difficult than the ones required for the assessment of potential hazards. IBAMA has technical personnel with extensive experience in the conduct and analysis of experiments and studies carried out in the laboratory or in the field. It also has access to data related to pesticides already tested in other countries. Since IBAMA currently does not have the resources to conduct risk assessments, it will continue its methods for evaluating potential hazards until better conditions permit the use of risk assessment.

It is possible that IBAMA, soon, can graduate from using the current methods for evaluation of potential hazard to performing environmental risk assessments associated with the use of pesticides. With Act # 84 (*3*) field studies or environmental monitoring may be required according to the risks and uses of the product. To achieve the use of environmental risk assessment for these products, IBAMA technicians are improving their knowledge by participating in courses, workshops, congresses and symposia.

Environmental Potential Hazards Classification and Evaluation Requirements

For the evaluation and classification of environmental potential hazards of pesticides the following information and studies must be submitted when applying for a product license: chemical and physical properties of the product, mobility, R_f adsorption-desorption, biodegradation in soils, photolysis, hydrolysis, acute and chronic toxicity tests in different non-target organisms (microorganisms, earthworms, aquatic plants, honeybees, fish, birds and mammals), as well as teratogenic, mutagenic and carcinogenic studies.

The generation of these studies and other information necessary to apply for a license may be carried out by one company, by a consortium of companies, or may be by a transfer/sale of data from one company to another. The studies and experiments may be conducted in Brazilian laboratories as well as in the laboratories of other countries. With the aim of having more reliable studies and experiments, IBAMA and INMETRO (National Institute of Metrology, Standardization and Industrial Quality) published Act #66 (*4*), July 1997, that established criteria for the licensing of national laboratories, as well as the recognition of foreign ones that conduct studies or experiments for the environmental evaluation of pesticides according to Good Laboratory Practices Standards (GLPS).

The evaluation of potential hazards to the environment made in Brazil results in the following final classification of the products: unacceptable product - product for which a license cannot be obtained; highly dangerous product - Class I; very dangerous product - Class II; dangerous product- Class III; and slightly dangerous product - Class IV.

Labeling

An important consideration in the environmental evaluation of pesticides is labeling. Labeling includes the label itself, plus additional information from the manufacturer about the product: restrictions on its use, hazards to users, etc. Labeling is done after the product has been evaluated by the regulatory agencies. The final classification of the product, as well as its characteristics or partial parameters that will result in its inclusion in Class I, must be stated on the label (for example, 'highly persistent in the environment'). According to the Brazilian Legislation, the sale and the usage of pesticides in fields can only be performed according to an agronomist's prescription.

Thus, labeling is instrumental in ensuring the correct management of the product by preventing harmful effects that might be caused by any pesticide used in the field. The label must state the environmental precautions, as well as other precautions to be taken, that must be followed by the users in order to decrease the possible adverse impact to the environment and to human populations.

Environmental Policy

Field studies and/or pesticide monitoring following the product use patterns will allow IBAMA to adopt environmental safety policies in accordance with the realities in Brazil. These studies also will help our government authorities to know what kind of actions they can demand from manufacturers, registrants and users in order to improve the management of the use of pesticides. These actions may define the continued approval or the cancellation of a license, or alterations to the use of the pesticide. All actions will be based on the information obtained from the environmental monitoring of these products.

Literature Cited

1. *Legislação Federal de Agrotóxicos e Afins*. Ministério da Agricultura, do Abastecimento e da Reforma Agrária; Departamento de Defesa e Inspeção Vegetal, 1995, pp 120.

2. *Manual de testes para avaliação da ecotoxicidade de agentes químicos*. IBAMA-DIRCOF - Departamento de Qualidade Ambiental, 1990, pp 351.

3. *Governo da República Federativa do Brasil (1996) Diário Oficial da União DOU 206/96,* Brasília. Portaria Normativa n° 84 de 23 de outubro de 1996.

4. *Governo da República Federativa do Brasil (1997) Diário Oficial da União DOU 125/97,* Brasília. Portaria Normativa n° 66 de 3 de julho de 1997.

Chapter 19

Brazilian and South American Pesticide Registration: The Industrial Perspective

Thaïs Reis Machado

Novartis Biociências SA, P.O. Box 50, CEP 06785–300 São Paulo, Brazil

Registration requirements in the Mercosur Region are a copy of the EU Directive. All Mercosur member countries will have to adjust internal requirements before January 2000. Agreement meetings are being held and the process is going on. Nevertheless, the registration procedure is still not equivalent in some countries. This paper discusses, in general, how this process occurs in some countries of South America and what GLP requirements are needed.

The agrochemical market in Latin America is growing rapidly and is one of the most encouraging in the world. Brazil and Argentina are expected to be the main parties in this growth. Brazil had the fifth largest market in the world in 1995, with a business volume around 1.8 billion dollars, compared to 6.4 billion dollars in the USA (*1*). In 1996, Brazil, alone, represented 46% of the South America Agrochemical market, about 65% of Mercosur sales, and this number will probably be increased by the year 2000 (*2*). These numbers show the importance of Brazil in Latin America and explain why the agrochemicals industry is concerned with eventual changes in legislation and in any upcoming new requirements in Brazil, since it will certainly influence all of Latin America.

A detailed analysis of the Brazilian, Argentinean and Chilean current requirements for pesticide registration shows that there are no great differences among them. The requirements above are also very similar to those of the EU Directive. These countries are already following the Mercosur guidelines and all member countries will have them fully implemented by January 2000. Andean Pact countries also are following Mercosur guidelines.

Mercosur is the South American Common Market, whose member countries, Brazil, Argentina, Uruguay and Paraguay are discussing rules of integration. Chile is not a member country but has participated in the meetings, as well as Andean Pact countries.

The main difficulties for pesticide registration in South America during the next few years will probably be generated by the evolution of laboratory accreditation programs based on Good Laboratory Practices (GLP) (*3*) and the differences in the registration process itself, from one Mercosur member country to another. Up to now, Brazil and especially the Brazilian Ministry of Environment, IBAMA, has headed up the implementation of the new requirements in the Mercosur resolutions.

Laboratory Accreditation Programs Based on GLP

The most recent stage of pesticide registration in Brazil began in 1989, upon the publication of Law 7802 (*4*), which became effective through Decree 98 816 in 1990. The new law requires new ecotoxicological and toxicologial studies, mainly for formulations. Products used in forestry need to be registered according to the new law, and field research with new active ingredients requires an "Experimental Use Permit" before registration for marketing, manufacturing and use in the country. Even for already registered products, this law requires registration renewal according to the new requirements to update all existing studies. Moreover, there is a demand that most of these studies are to be conducted in Brazil, and data for registration are assessed by three ministries: IBAMA (*5*), Ministry of Health (*6*), and Ministry of Agriculture (MAA) (*7*).

Because of the large amount of data to be generated, most of the industry found it inconvenient to have the studies conducted in house; therefore, they contracted third party laboratories. Only a few multinational companies maintained their investment in some specialized laboratories, for example, pesticide residue analysis laboratories. As a result, many third party laboratories qualified themselves rapidly and began generating the necessary data. Unfortunately, Brazilian laboratories did not have much experience in conducting most of these studies.

Obviously, both the government and the industry realized that a quality monitoring system was necessary, since these studies are strategic, very expensive and must be reliable. Consequently, the concept of Good Laboratory Practices (GLP) started being discussed and was eventually adopted. On the other hand, competition among laboratories led to many discussions related to scientific quality, method validation and laboratory accreditation. Meetings were held that involved industry, public and private laboratories, authorities, and universities.

The development of these discussions culminated in an agreement between IBAMA and the National Institute of Metrology, Standardization and Industrial Quality, INMETRO. Today, laboratories generating data submitted to IBAMA for registration must have an INMETRO accreditation which they receive after being audited for compliance with the GLP Principles (*8*).

Response of the Laboratories. These criteria, adopted to meet accreditation standards, have caused much discussion. Arguments and doubts put forth were related to interpretation of the criteria and other considerations, such as, the deadline for GLP compliance, the selection of auditors, costs, and the way audits should be conducted.

Deadline for GLP Compliance. It is known that laboratories usually take a long time to achieve full GLP compliance. Because of this, IBAMA and INMETRO accreditation is voluntary and is composed of three steps, with each step increasing the level of compliance. Laboratories can start in at any step and can have studies accepted after they have passed the first step.

Selection of Auditors. Audits will be conducted by an INMETRO auditor and a consultant. Some consultants are working with third party laboratories and it could generate problems between competitors. However, laboratories can refuse to admit a consultant if they believe there is a conflict of interest.

Acceptance of Data Generated In House and International Acceptance of Brazilian Data. How would it be discussed, considering the lack of a Memorandum of Understanding? Many uncertainties and questions have arisen between Brazil and other countries about procedures since countries like the US do not require laboratory accreditation.

Increase in Cost. It is known that adoption of GLP increases a laboratory's cost by 30%.

The Industry Problem. Industry is especially concerned because it relies mainly on third party laboratories to conduct these studies. Even with all these discussions, there was consensus that GLP is an effective and needed instrument. Today, most of the Brazilian laboratories involved in studies to be conducted for registration have already started implementing GLP. Although no audit has been performed yet by INMETRO, the program is coming along and working under GLP is no longer questioned.

Mistakes are expected to happen because our society is in a learning mode with this process. Certainly a great deal of discussion is required between the authorities and the laboratories, as has happened in countries where GLP has been adopted. These discussions must be conducted in a spirit of collaboration and trust so that the accreditation process is accomplished swiftly and successfully. Radical positions and quick conclusions should be avoided in order to bring us data quality and reliability.

Society of Quality Assurance. A Society of Quality Assurance, working along the same lines as the American Society of Quality Assurance, is desirable, with authorities and QA professionals involved, meeting face to face and discussing their problems scientifically. This should be particularly helpful for Brazilian Quality Assurance professionals and for consultants and government auditors because they would have a chance to discuss and compare procedures with the more experienced professionals in the field. This Society should be supported by laboratories, universities, government authorities and industry, without political influences and from a neutral aspect, in order to learn about and debate issues so that the decisions made serve to reach the desired harmonization and reliability.

Effects. With regard to the IBAMA-INMETRO agreements and Joint Decree (*9*), Brazilian laboratories have had to deal with both positive effects and numerous difficulties, not too different really from the ones faced by American and European laboratories.

Difficulties.

- Entitle the right persons as Study Directors and Quality Assurance staff,
- Close monitoring of costs, in order to avoid unnecessary increases due to excessive bureaucracy,
- Develop and implement numerous required Standard Operating Procedures,
- Convince management that GLP is not an overload of documentation, and
- Economic resources available for the maintenance and calibration of equipment and supply of chemicals, mainly in public laboratories.

Positive Effects.

- Suppliers of services and manufacturers of glassware, chemicals and equipment are improving quality of materials supplied with ISO certification or INMETRO accreditation,
- Laboratory professionals are more flexible and support the implementation of GLP,
- Organization of facilities has been improved,
- Studies are better planned with fewer false starts,
- Staff are better trained, and
- GLP standards are more and more accepted and adopted.

Differences in the Processes of Pesticide Registration in Mercosur Member Countries

Difficulties faced by the industry, regarding pesticide registration itself, are due to the expensive and time consuming registration process. Although the average time taken to approve new active ingredients is expected to be reduced in Brazil, registration fees are high.

Registration data in Brazil are submitted to the Ministry of Agriculture, whereas toxicological data are evaluated by the Ministry of Health, and ecotoxicological data are evaluated by IBAMA. Accreditation is given by INMETRO.

Moreover, the State of Paraná, in Brazil, like State of California, in US, has its own legislation, which influences other states such as Minas Gerais and Espírito Santo. Paraná's legislation (*10*) stipulates that residue data must be generated by public laboratories; however, it does not demand that such laboratories offer a quality program.

In Argentina (*11*), data for pesticide registration are submitted to the Ministry of Agriculture and are evaluated by consultants. GLP is not required yet. "Experimental Use Permits" are required also in that country. A new law is coming up, with new rules facilitating the registration of generic pesticides or products with similar active ingredients. Chile (*12-13*) is following the Mercosur resolutions; however, Uruguay and Paraguay are not harmonized yet with the Mercosur guidelines.

Resolution 73/94 of Mercosur (*14*) establishes that "when international rules or standards in the field of quality of technical products and formulations are available, they shall be adopted by member states after the respective evaluation by the Sanitary Committee. When the same rules are not available or are not adequate to meet regional requirements, the Sanitary Committee will establish new rules and standards to be approved by the Common Market Council".

OECD (*15*) states that "Principles of Good Laboratory Practices should be applied to testing of chemicals to obtain data on their properties and/or their safety with respect to human health or the environment." "Comparable quality of test data forms the basis for the mutual acceptance of test data among countries...duplicative testing can be avoided, thereby introducing economies in test costs and time. The application of these Principles should help avoid the creation of technical barriers to trade and further improve the protection of human health and the environment."

Certainly, these principles are one of the most adequate instruments to produce reliable data, and all studies for registration purposes in South America (Tables I, II, III and IV) will probably be required to be conducted under GLP compliance.

Conclusions

Industry believes governments will become more and more demanding for data concerning the toxicity of active ingredients.

Industries operating in Latin America certainly will have to deal with issues concerning registration, audits, accreditation, and GLP implementation. We have all learned throughout this process and some misunderstandings are to be expected. The need for a Society as a forum to discuss issues should be a priority.

Testing facility management must actively support GLP and Quality Assurance Programs in their facilities.

Authorities should strongly support their personnel involved in data assessment for registration and provide them with all the necessary training in order to reach the final target of human and environmental protection with efficiency. Industry is actively supporting these training opportunities.

Literature Cited

1. *Agrow* **Sep 12th, 1997**, *228*, 1.
2. Trade of Phytosanitary Products in Brazil and Argentina, *Sumário executivo* **May 1997,** RCW Consultores, Research International Brazil and Mora y Araujo Nogueira Associados.
3. Good Laboratory Practice Standards – *Code of Federal Register, Part 160, FR doc.89-19087* Filed **8-16-89.**
4. Brazilian Law 7802 from Jul 11, 1989, *Diário Oficial da União*, **Jul 12th, 1989**.
5. Decree 84, IBAMA , *Diário Oficial da União*, **Oct 23, 1996**, *206*.

Table I. Testing Requirements for an Experimental Use Permit - Argentina.

Experimental Use Permit – Argentina

AI=active Ingredient

FP=Formulation

Formulation		
TEST	**Tested**	**Note**
Physico-Chemistry Data		
(C1) - Physical State, Aspect, Color and Smell	AI	Report
(C2) - Structure and Molecular Formula	AI	Statement
Detailed Composition	FP	Statement
TOXICITY:		
Acute Oral - Rats	AI	Report
Acute Dermal - Rats	AI	Report
Inhalation - Rats	AI	Report
Skin Irritation	AI	Report
Eye Irritation	AI	Report
Skin Sensitivity	AI	Report
Mutagenicity - Ames (at least)	AI	Report
28-Day Oral (rats)	AI	Report
90-Day Oral Toxicity - Dogs and Rodents	AI	Report

Table II. Testing Requirements for an Experimental Use Permit - Brazil.

Experimental Use Permit - Brazil

TG=Technical Grade

CR=Conditional requirement

Formulated Product		
TEST	tested	Note
Physico-Chemistry Data		
(C1) - Physical State, Aspect, Color and Smell	AI	Report
	FP	Report
(C2) - Structure and Molecular Formula	AI	Statement
Detailed Composition	FP	Statement
(C8) - Solubility/Miscibility in Water and Other Solvents	AI	Report
(C9) - pH	FP	Statement
(C14) - Partition Coefficient n-Octanol/Water	AI	Report
(C15) - Density or Relative Density at 20°C in g/l (Liquid Formulation)	AI	Report
	FP	Report
Wetability (Dispersible Powders)		CR - Report
TOXICITY:		
Acute Oral - Rats	AI	Report
	FP	Report
Acute Dermal - Rats	AI	Report
	FP	Report
Inhalation - Rats	AI	Report
Skin Irritation	AI	Report
	FP	Report
Eye Irritation	AI	Report
Skin Sensitization	AI	Report
Mutagenicity - Ames	AI	Report
Mutagenicity - Micronucleus	AI	Report
Toxicological Compatibility (Ready Mixtures)		CR - Ready Mixtures
Sub-Acute Oral Toxicity - Dogs and Rodents	AI	Report
Long Term Toxicity - Mammals	AI	Report
Medical Information		
Intoxication Symptoms and Diagnosis	FP	Report
Intoxication Symptoms and Diagnosis	AI	Report
Recommended Medical Treatment		Report
First Aid	FP	Report
Antidote	FP	Report

Table II. Continued.

Toxic Effects in Other Species		
Birds - Acute Oral	AI	Report
Birds - Dietary		Report
Fishes - Acute	AI	Report
Daphnia - Acute	AI	Report
Effects on Algae	AI	Report
Phytotoxicity to Nontarget Plants		Report
Bee Toxicity	AI	Report
Earthworm Toxicity	AI	Report
Soil Microorganism Toxicity		Report
Beneficial Arthropods (Predators)		Report
Environmental Fate		
Behavior in Soil (3 Standard Soils)		If Pertinent
Leaching	AI	If Pertinent
Degradation	AI	If Pertinent
Behavior in Water and Air (Residues and Degradation)	AI	If Pertinent
Volatility	AI	If Pertinent
Behavior in Water and Air (Residues and Degradation), Biodegradation, Hydrolysis and Photolysis	AI	If Pertinent
Residue Data		
Local Residue Data - 2 Seasons		Report
Residue Data Produced According to FAO Recommendation for Residue Trials		Report
Safety Measures		
A.I. Destruction and Decontamination Procedures	AI	Information
Recovery Potential		Information
Neutralization Potential		Information
Controlled Incineration		Information
Water Depuration		Information
Identity of Combustion Products		Information
Application Equipment Cleaning Procedures		Information
Safety Precautions: Handling, Storage, Transport, Accidental Fire or Spillage	AI	Information
Individual Protection Equipment		Information
Registration in Other Countries		
Country of Origin	AI	Information
USA/Europe/Japan	AI	Information
COSAVE/MERCOSUL Countries	AI	Information

Table III. Testing Requirements for a Technical Product in Argentina, Brazil, and Chile.

TECHNICAL PRODUCT: **A-Argentina, B-Brazil, C-Chile** AI=Active Ingredient TG=Tecnical Grade		
Requirement	**Test**	**Country**
(C1) - Physical State, Aspect, Color and Smell	AI	A,B,C
Structure and Nomenclature	AI	A,B,C
(C2) - Molecular Identification/ Absorption Spectra	AI	A,B,C
(C3) - Purity	AI	A,B,C
Detailed Composition	AI	A,B,C
> 0.1% Impurities		A,B,C
Impurities - Tox. Significance	AI	A,B,C
Manufacturing Scheme	AI	B
Name/Address of Producer/Raw Material Suppliers	AI	B
(C4 -)Metallic Impurities	AI	B
(C5) - Melting Point (Range)	AI	A,B,C
(C6) - Boiling Point (Range)	AI	A,B,C
(C7) - Vapor Pressure	AI	A,B,C
(C8) - Solubility in Water and Other Solvents	AI	A,B,C
(C9) - pH	AI	A,B,C
(C10) - Dissociation Constants in Water	AI	A,B,C
(C11) - Complex Formation Ability in Water	AI	B
(C12) - Hydrolysis	AI	A,B,C
(C13) - Photolysis	AI	A,B,C
(C14) - Partition Coeficient - n-Octanol/Water	AI	A,B,C
(C15) - Density	AI	A,B,C
(C16) - Superficial Tension of Solutions	AI	A,B,C

Table III. Continued.		
(C17) - Viscosity	AI	A,B,C
(C18) - Particle Size Distribution	AI	A,B,C
(C19) - Corrosivity/Oxidizing Properties	AI	A,B,C
(C20) - Thermal and Air Stability	AI	A,B,C
Volatility	AI	A,C,B
Flammability	AI	A,C
Explosive Properties	AI	A,C
Reaction to the Packaging Materal	AI	A,C
Analytical Methods	AI	A,B,C
Analytical Methods	Isomers., Impurities Degradat. Products, Additives	A,B,C
Analytical Methods	Residues (Parent & Metabolite s), in Plants, Food, Soil and Water	A,B,C
Analytical Methods	Air, Animal and Human Tissues and Fluids	C
Plant Metabolism	AI	A,B,C
Identification of Degradation Products and Metabolites	AI	A,B,C
Degradation Pattern in Plants	AI	A,B,C
Residue Data	FP	A,B,C
Acute Oral - Rats	AI	A,B,C
Acute Dermal - Rats	AI	A,B,C
Inhalation - Rats	AI	A,B,C
Skin Irritation	AI	A,B,C
Eye Irritation	AI	A,B,C
Skin Sensitization	AI	A,B,C
Oral - 28 Days	AI	A,B,C
Oral - 90 Days (Two Species)	AI	A,B,C

Continued on next page.

Table III. Continued.		
Fishes- Effects in Reproduction and Growth	AI	A,C
Fishes - Bioacumulation	AI	A,B,C
Daphnia - Acute	AI	A,B,C
Daphnia - Chronic	AI	A,B,C
Daphnia - Reproduction and Growth	AI	A,B,C
Effects on Algae	AI	A,B,C
Phytotoxicity	AI	A,B,C
Bee Toxicity (Oral and Contact)	AI	A,B,C
Bee - Mortality - Field Simulation	AI	A,C, B
Earthworm Toxicity	AI	A,B,C
Soil Microorganism Toxicity	AI	A,B,C
Beneficial Arthropods (Predators)	AI	A,C
Behavior in Soil (3 Standard Soils)	AI	A,B,C
Degradation in Soil	AI	A,B,C
Adsorption/Desorption in Soil	AI	A,B,C
Biodegradation in Soil	AI	A,B,C
Ready Biodegradability	AI	B
Leaching Behavior	AI	A,B,C
Behavior in Water and Air (Residues and Degradation)	AI	A,B,C
Degradation in Aquatic Environments	AI	A,B,C
Hydrolysis	AI	A,B,C
Photolysis	AI	A,B,C
Parent Compound and Metabolites Residue Analysis from Residue Trials According to GFP	Formulation	A,B,C
Local Residue Data - 2 Seasons	Formulation	A,B
A.I. Destruction and Decontamination Procedures.	AI	A,B,C
Recovery Potential	AI	A,B,C
Neutralization Potential	AI	A,B,C
Controlled Incineration	AI	A,B,C
Water Depuration	AI	A,B,C

Table III. Continued.		
Dermal	AI	A,B,C
Inhalation	AI	A,B,C
Oral - 1 Year Dogs	AI	A,B,C
Oral Chronic Tox. - 2 Years (2 Species)	AI	A,B,C
Carcinogenicity	AI	A,B,C
Mutagenicity (Procaryotes/Eucaryotes)	AI	A,B,C
Teratogenicity - 2 Species	AI	A,B,C
Reproduction Effects	AI	A,B,C
Toxicological Compatibility (Ready Mixtures)	AI	A,B,C
Metabolism in Mammals - Absorption, Distribution, Excretion, Metabolic Pathways)	AI	A,B,C
Intoxication Symptoms and Diagnosis	AI	A,B,C
Recommended Medical Treatment	AI	A,B,C
First Aid	AI	A,B,C
Antidote	AI	A,B,C
Neurotoxicity	AI	A,B,C
Toxicity of Metabolites	AI	A,B,C
Intoxication Diagnosis: Direct Observation in Clinical Cases	AI	A,B,C
Health Control Registers in Industry and Use in Other Countries	AI	A,B,C
Health Control Registers-Experimental Products Use	AI	A,B,C
Health Control Registers-Commercial Use	AI	A,B,C
Epidemiological Studies/Population Data	AI	A,B,C
Sensitization Data	AI	A,B,C
Birds - Acute Oral	AI	AB,C
Birds - Diet	AI	AB,C
Birds - Reproduction	AI	A,C,B
Fishes - Acute	AI	A,B,C
Fishes - Chronic	AI	A,B,C

Continued on next page.

Table III. Continued.		
Safety Precautions: Handling, Storage, Transport, Accidental Fire or Spillage.	AI	A,B,C
Individual Protection Equipment	AI	A,B,C
Combustion Products in Case of Fire	AI	A,C
Country of Origin	AI / FP	A,B,C
USA/Europe/Japan	AI / FP	A,B,C
COSAVE/MERCOSUR Countries	AI / FP	A,B,C

Table IV. Testing Requirements for a Formulated Product.

FORMULATED PRODUCT:				A NUMBER:
ACTIVE INGREDIENT:				
Requirement	**TEST**	**Guideline**	**Country**	**Note**
Physico-Chemistry Data				
(C1) - Physical aspect:	Formulation	not pertinent	A,B,C	Study
(C1) - Physical state	Formulation	FIFRA 63-3	A,B,C	Study
(C1)-color	Formulation	FIFRA 63-2	A,B,C	Study
(C1)-smell	Formulation	FIFRA 63-4	A,B,C	Study
Detailed Composition	Formulation	not defined	A,B,C	Statement and Study
Degradation Products	Formulation	not defined	A,B,C	Statement
(C8) - Miscibility in water and other solvents	Formulation	OECD 109; CIPAC MT 46; CIPAC MT 39	A,B,C	Report
(C9) - pH	Formulation	CIPAC MT 75 /CIPACMT31/ FIFRA 63-12	A,B,C	Statement
(C15) - Density or relative density at 20ºC in g/l (liquid or solid formulations at RT)	Formulation	OECD 109/FIFRA 63-7; A3	A B, C	Study
(C16) - Superficial Tension of solutions	Formulation	OECD 115	A,B,C	Study
(C17) - Viscosity (liquids at RT)	Formulation	CIPAC MT 22;OECD 114,oil,suspensions&emulsions /FIFRA 63-18	A,B,C	CR - Study
(C18) - Particle Size Distribution - Solids at RT	Formulation	OECD 110	B, A	Study
(C19) - Corrosivity	Formulation		A,B,C	Study
(C20) - Thermal and Air stability	Formulation	OECD 109?	A,B,C	Study
Shelf life	Formulation	CIPAC MT 39; MT 46; OECD 109; FIFRA 63-17	A, C, B	Study
Volatility	Formulation	FIFRA 163-2	A,C	Study
Ponto Fulgor/Flamability	Formulation	FIFRA 63-15	A,C,B	CR
Inflamability Point		FIFRA 63-15	A,C	CR oil&solutions
Explosive Properties	Formulation	FIFRA 63-16	A, B, C	CR - A14
Wetability(dispersible powders)	Formulation	CIPAC MT 53.3	A,C	CR - Dispersible Powders

Continued on next page.

Table IV. Continued.

Requirement	TEST	Guideline	Country	Note
Foam Persistence	Formulation	CIPAC MT 47	A,C	CR - application with water
Suspensibility	Formulation	CIPAC MT 15 (WP); CIPAC MT 160/161 (FW); CIPAC MT 168/174 (WDG)	A,C	only for dispersible powders & fws
Wet sieve	Formulation	CIPAC MT 59,3, MT 167 (WDG)	A,C	CR- dispersible powders & SCW)
Dry sieve	Formulation	CIPAC MT 59,1(P); CIPAC MT 58 (G)	A,C	CR- granules&powders
Emulsion stability	Formulation	CIPAC MT 20/MT 36 (visual *); CIPAC MT 173 (photometric **)	A,C	CR-EC: *- Test conc. 5%;** Test conc < = 2%
Incompatibility	Formulation	not defined	A,C	CR-EC
Sulfonation index	Formulation	not defined	A,C	CR-oil
Dispersion	Formulation	not defined	A,C	CR- dispersible granules
Gases Production	Formulation	not defined	A,C	Statement
Fluidity	Formulation	CIPAC MT 172	A,C	CR-dry powders
Iodine index &saponification index	Formulation	not defined	A,C	CR - oil
Packaging Profile	Formulation	complete profile	A,B,C	information
Reaction to the packaging material	Formulation	not defined	A,B,C	report
Disposal of Empty Pesticides Containers	Formulation	not defined	A,B,C	information
Manufacture Process - summary	Formulation	not defined	B	statement
Name address of Formulation components suppliers	Formulation	not pertinent	B	Information
ANALYTICAL METHODS:				
ANALYTICAL METHODS	Formulation	not pertinent	A,B,C	

Table IV. Continued.

Requirement	TEST	Guideline	Country	Note
ANALYTICAL METHODS	residues (parent & metabolites)	not pertinent	A,B,C	
TOXICITY:				
Acute oral - rats - F1.1.1	formulation	OECD 401/420	A,B,C	Study - FIFRA 81-1
Acute dermal - rats - F 3.1	formulation	OECD 402	A,B,C	Study - FIFRA 81-2
Inhalation - rats - F 2	formulation	OECD 403	A,B,C	CR - Study - FIFRA 81-3
Skin irritation - F 3.4	formulation	OECD 404	A,B,C	Study-FIFRA 81-4
Eye irritation - F 4	formulation	OECD 405	A,B,C	Study-FIFRA 81-5
Skin sensitizing -	formulation	OECD 406	A,B,C	Study-FIFRA 81-6
Mutagenicity - (procaryotes/eucaryotes) - G.1.1.1/G.1.1.2/G.1.1.3/G.1.2.1/ G.1.2.2/G.1.2.3/G.1.2.4/G.1.2.5-	formulation	VERSION 87	B	Study - Always required in Brazil
Toxicological compatibility (ready mixtures)	formulation	not defined	A,B,C	CR - ready mixtures
Medical Information				
Intoxication symptoms and diagnosis	formulation	not pertinent	A,B,C	Report
Recommended Medical Treatment	formulation	not pertinent	A,B,C	Report
First Aids	formulation	not pertinent	A,B,C	Report
Antidote	formulation	not pertinent	A,B,C	Report
Additional studies:				
Neurotoxicity	formulation	not defined	B	CR
Complementary medical information:				
Intoxication Diagnosis:direct observation in clinical cases	formulation	not pertinent	A,B,C	CR
Health Control Registers in Industry and use in other countries	formulation	not pertinent	A,B,C	if available
Health Control Registers-Experimental products use	formulation	not pertinent	A,B,C	if available
Health Control Registers-Commercial Use	formulation	not pertinent	A,B,C	if available
Epidemiological studies/Population data	formulation	not pertinent	A,B,C	if available
Sensitization data	formulation	not pertinent	A,B,C	if available
Re-entry interval	formulation		A,B,C	information

Continued on next page.

Table IV. Continued.

Requirement	TEST	Guideline	Country	Note
TOXIC EFFECTS IN OTHER SPECIES:				
(D 8.1) Birds - Acute oral	formulation	FIFRA 71-1	A,B,C	Study
Birds - dietary and Reproduction	formulation	not defined	A, B	Study
(D 6.1)Fishes - acute & chronic	formulation	Acute: OECD 203;FIFRA 72-1	A,B,C	Study-NOTE:chronic studies are not performed with formulations-justification is accepted
Daphnia - acute - D 5.1	formulation	OECD 202	A,B	Study-FIFRA 72-2;ISO 6341
Effects on algae - D 2	formulation	OECD 201	A,B	Study-FIFRA 123-2;ISO 8692
Phytotoxicity to non target plants D 9.1	formulation	not defined	A,B,C	Study/CR
Bees Toxicity - D 4	formulation	EPPO 170	A,B,C	Report
Earthworms Toxicity D 3	formulation	OECD 207	B,C	Report
Soil microorganisms toxicity - D3	formulation	OECD	B,C	Report
Beneficial arthropods (predators)	formulation	not defined	C	Report
Environmental fate				
Behaviour in Soil (3 standard soils)	formulation	not defined	A,C	if pertinent-justification is accepted
Leaching	formulation	not defined	A,C	if pertinent-justification is accepted
Degradation	formulation	not defined	A,C	if pertinent-justification is accepted
Behaviour in water and air (residues and degradation)	formulation	not defined	A,C	if pertinent-justification is accepted
Behaviour in water and air (residues and degradation), biodegradation, hidrolysis and photolysis	formulation	not defined	C	if pertinent-justification is accepted
RESIDUE DATA				
Local residue data - 2 seasons or 2 regions	formulation	FAO	A,B	Study

Table IV. Continued.

Requirement	TEST	Guideline	Country	Note
Residue Data produced according to FAO recommendation for Residue Trials	formulation	FAO	A,B,C	Report
Safety Measures				
A.I. Destruction and decontamination procedures.	formulation	not defined	A,B,C	Information
Recovery Potential	formulation	not defined	A,B,C	Information
Neutralization potential	formulation	not defined	A,B,C	Information
Controlled incineration	formulation	not defined	A,B,C	detailed information
Water depuration	formulation	not defined	A,B,C	Information
Identity of combustion gases in the case of fire	formulation	not defined	A,C	Statement
Application Equipment cleaning procedures	formulation	not defined	A,B,C	Information
Safety Precautions:handling, storage, transport accidental fire or spillage.	formulation	not defined	A,B,C	Information
Individual Protection Equipment	formulation	not defined	A,B,C	Information
Registration in other countries				
Country of origin	formulation	not pertinent	A,B,C	Information
USA/Europe/Japan	formulation	not pertinent	A,B,C	Information
COSAVE/MERCOSUR countries	formulation	not pertinent	A,B,C	Information
MRL´s:				
Novartis´s proposal	not pertinent	not defined	A,B,C	Information
Other countries	not pertinent	not defined	A,B,C	Information
Limit of unknown residues	not pertinent	not defined	B	Information
Residues in processed food	not pertinent	not defined	B	Information
Residues in ração	not pertinent	not defined	B	Information
Intervalo de Reentrada	not pertinent	not defined	B	Information
A= ARGENTINA ; B= BRAZIL ; C= CHILE ; CR = Conditional requirement ; Report - complete report is needed		**Temporarily Special Registration: Only in Brazil and Argentina**		

6. Decree 03 de Jan 16, 1992, Secretaria Nacional de Vigilância Sanitária, Guideline Nr 01 Dec 9, 1991 of the Law 7802, regulated by the Decree 98816 from Jan 11, 1990, *Diário Oficial da União*, **Feb 04, 1992.**
7. Decree 45, from Dec 10, 1990, *Diário Oficial da União,* **Dec 14,1990**.
8. Princípios de Boas Práticas de Laboratório, *INMETRO*, **1995**.
9. Joint Decree 66, IBAMA and INMETRO, *Diário Oficial da União,* **Jul 03 1997**, *125*.
10. Paraná : Resolution 24/90, *Diário Oficial do Estado do Paraná*, **Apr 24, 1990**.
11. Argentina: Resolution 17/95, Manual de Procedimientos para Registro de Productos Fitossanitarios en Argentina, *Boletim Oficial*, **Mar 21, 1995**.
12. Chile: Resolución 1178 del Servicio Agrícola e Ganadero, *Diário Oficial da União*, **Aug 24,1984.**
13. Law18755 – Lei Orgânica del Servício in cooperation with FAO.
14. Resolution 73/94 , Mercosur Commom Market Council, in the view of the Asunción Treaty, Art 10,of the decision 4/91 and decision 1/93 of the CMC and Recommendations 7/94 and 15/94 of SGT, Agriculture Policy.
15. The OECD Principles on Good Laboratory Practice, OECD series on Principles of Good Laboratory Practice and Compliance Monitoring, Nr 01**.**, *Environment Directorate, OECD*, **1998**.

Chapter 20

Status of Pesticides Control in Cordoba, Argentina

Mirtha Nassetta and Sara Palacios

Center of Excellence on Products and Processes of Cordoba, Alvarez de Arenales 230, (5000) Cordoba, Argentina

Federal regulatory requirements for pesticides in Argentina are under the supervision of SENASA (The Official Argentinean Institution for the health and quality control of crops, vegetables and livestock derivatives).

CEPROCOR (Center of Excellence on Products and Processes of Córdoba) is a novel research center under the auspices of the Government of the Province of Córdoba. CEPROCOR's mission is cooperative research to meet industrial and social needs, including the provision of human resources at the post-graduate level. CEPROCOR's current projects include analytical chemistry, fine chemicals, biosciences, thermal dosimeter and remote sensors, and some undertaken in agreement with local industries.

CEPROCOR currently performs pesticide residue determinations in water, soil, food, and biological samples. We have been certified recently by SENASA. One of our main interests is to improve the quality of our analytical procedures and set up new strategies and techniques in order to satisfy present environmental and clinical demands in our region.

History of CEPROCOR

Cordoba is a province in the republic of Argentina and is located in the center of the country. It is one of the largest cities on the commercial route from the Pacific Ocean to the Atlantic Ocean within the Common Market of South America (MERCOSUR) formed by Argentina, Brazil, Uruguay and Paraguay. The economic activity of Córdoba is concentrated in three areas:

- Agriculture, cattle raising, fishery and forestry (24% Gross Provincial Product (GPP)).
- Manufacturing industry (19% GPP).
- Commerce and services (30% GPP).

The main industrial areas in Córdoba are metallurgy and food processing. In this context, CEPROCOR (Center for Excellence in Products and Processes of Córdoba) was founded under the area of the Secretary of Science and Technology of the Government of Córdoba. The main objective of CEPROCOR is to incorporate Science and Technology into the production of goods and services within the private or public sector, with the aim of assuring its efficiency and competitiveness in the national and international markets.

In 1995, CEPROCOR began its activities in research and development full time. Since then, the work of the Center has been oriented toward two fundamental areas:

- Analysis and quality control.
- Research and development.

In the Area of Quality Control, CEPROCOR is at the moment fully dedicated to the improvement of its laboratories in different areas, such as, pesticides, water and the environment, the food supply, pharmaceutical compounds, molecular virology, and DNA fingerprinting.

In the area of research and development, the Center is carrying out projects related to new natural pesticides, methods for the purification of biomaterials, diagnostic kits, new analytic techniques for the control of pharmaceutical compounds, synthesis of quirals compounds used as medicaments, expression of recombinant proteins, molecular and atomic spectroscopy, and a dosimeter for radiation.

Pesticide Residue Laboratory: History and Perspective

The Pesticide Residue Laboratory (PRL) of CEPROCOR was born in 1995. This was a necessity for the peanut and potatoe producers of the Province of Córdoba (the province of Córdoba is the principle producer of peanuts in the country, 98 %), who needed to certify the quality and health of their products for export. Until that time, this certification was carried out only by SENASA (The Argentinean Institute for the Health and Quality Control of crops, vegetables, and livestock derivatives), which is located in the Capital of the country, 600 miles away from the production areas.

About the same time (and after the Laboratory became known in the city of Córdoba), we began to receive several requests from the largest supermarket chain in the city to determine the pesticide residues in the food that they sell.

We also received requests from small producers of fruit and vegetables, and we were requested to determine pesticide residues in water for human consumption and for irrigation, in soil from the most important peanut and soybean producing areas in the Province, in milk by-products, and in honey for export.

The most frequently detected pesticides were: chlordane in river water; DDT and its metabolites in well water; and lindane in soil. When organophosphorus pesticides were analyzed, we detected methyl chlorpyrifos in cheese for export, ethion in grapefruit, DDVP in green pumpkins, and others, such as, deltamethrin in cheese and trifluralin in soil. We frequently analyze blood from workers who were exposed to pesticides for a long time. We conduct these analyses by measuring the value of the acetyl cholinesterase and the pesticide level in blood and urine.

In January and February 1997, we had several cases of children that had become intoxicated from organophosphorus pesticides (some of the children were poisoned with parathion, a pesticide whose use has been banned since 1994). In all cases, the agent causing the intoxication was identified; therefore, a quick medical treatment was possible.

We also conduct analyses to determine residues of PCB in the soil and in well water. These studies are required by the City before the start up of any new industrial project. We did not find PCB residues in the samples analyzed. Some of these results are summarized in Table I.

To conduct all these studies we use the techniques from the Association of Official Analytical Chemists, the Pesticides Analytical Manual edited by Food and Drug Administration and, if it is necessary, we use new techniques previously validated for us that compare the results by spiked samples and measured recoveries.

Cattle raising is also a very important activity in our Province. The meat packing houses export meat to the European Common Market, the United States, and other countries. All of these countries require the meat to be checked for organochloride pesticide residues, such as, lindane, aldrin, HCH, HCB, heptachlor, etc., in the place of origin of the products.

The national organization that certifies the laboratories that are able to do these studies is SENASA. They have a program of inspections to assure the capability of the laboratories which conduct the required analyses. To become certified, a laboratory must first register and pay a fee of US $400 per year. An inspector then comes to the laboratory and conducts a thorough facility inspection.

Each chemist working under this certification, must perform a proficiency test on a 'control' twice a year. Controls include proficiency samples spiked with unknown organochlorides and organophosphorus residues at different concentration levels. The inspectors evaluate the analytical performance of the laboratory. If the laboratory fails, it can try again. If it fails again on the second try, it does not get certified or it loses its certification if it had one previously. If the laboratory passes, a certificate is issued and is good for one year. The following year, the laboratory must be recertified and follow the same procedure.

There is a quality assurance type of manual that describes what a laboratory must do to be a certified. It contains sections on laboratory organization, management and operating personnel, procedures for maintaining and testing laboratory equipment, criteria for the selection of testing methods and procedures for analytical method validation, sample handling and storage and disposal of stock solutions, requirements for reporting and validating analytical results, and audit or control procedures. These inspectors are 'the law' and laboratory personnel must

Table I: Results of Sample Analysis for Pesticide Residues.

SAMPLES	PESTICIDES	SAMPLES WITH NO RESIDUES FOUND	SAMPLES VIOLATIVES Over Tolerance (a)	SAMPLES VIOLATIVES Under Tolerance (a)
1. River water	Chlordane		Yes	
2. Drinking water (Córdoba City)		Yes		
3. Well water	PCB	Yes		
4. Well water	DDT (metabolites)		Yes	
5. Soil	PCB	Yes		
6. Soil	Lindane		Yes	
7. Soil	Trifluralin		Yes	
8. Cheese	Me-Chlorpyrifos		Yes	
9. Cheese	Deltamethrin			Yes
10. Chicken feed	Fenitrothion		Yes	
11. Grapefruit	Ethion		Yes	
12. Green pumpkins	DDVP		Yes	
13. Honey	Fenitrothion		Yes	
14. Essential oils		Yes		

(a) Federal Regulatory Limits

do as the inspector wishes to assure certification. Our Pesticide Residue Laboratory belongs to the network of Official Laboratories of SENASA .

Moreover, to assure quality of the results of our analyses, we are starting a collaborative program of the GTZ (Pesticide Service Project) - World Health Organization (Collaborating Center for Pesticide Analysis and Training) related to an Analytical Quality Assurance Study. At this moment, we are carrying out determinations of pesticide residues on rice flour that were provided by the GTZ-WHO program.

The Secretary of Environmental Protection is planning to monitor monocrotophos residues in the south of our Province. This is because a significant decrease in the population of young eagles was observed. This situation causes many ecological disorders, including an increased threat from grasshoppers. Our Pesticide Residue Laboratory will collaborate in this study that measures the level of monocrotophos in soil, water, crops from the region, and also in young eagle blood .

Our Pesticide Residues Laboratory is one of only three Laboratories of this kind in our Country. One of them is located in the Patagonia and the other is located in the Capital.

We hope that with economic and technical cooperation by national and international organizations we can carry out more effective pesticide residue monitoring. These activities will include work with government and industry sectors to establish the control of pesticide usage to better measure compliance of our exports with other countries' tolerances. This will also improve the economic development of certain regions in the northeast and northwest of our Country.

Chapter 21

International Multi-Country Field Studies: GLP Problems and Solutions

Markus M. Jensen

Jenerations Consulting, Inc., 565 Petite Prairie Road, Washington, LA 70589

Conducting international multi-country field studies is challenging as well as complex. U.S. EPA FIFRA and OECD Good Laboratory Practice (GLP) studies require detailed planning, training, implementation, and documentation among numerous individuals of various cultural and scientific disciplines. Required quality assurance oversight and responsibilities are equally challenging as these studies increase in magnitude and complexity. Major considerations from the sponsors' and, when applicable, contractors' points of view are described regarding these multi-continent, multi-country study scenarios. An examination of some of the major analytical and field problems and concerns along with related solutions are presented and discussed.

Good Laboratory Practice (GLP) field research is not a predictable science from one year to the next because of changing environmental conditions, varying test systems, cultural practices and protocol requirements. Conducting international multi-country field studies presents additional challenges to this complex research. It is difficult to attempt to address all of the wide range of issues involved in these studies; therefore, some of the major considerations from the sponsor's point of view as well as key study and communication problems with related solutions will be addressed.

I work with an independent field contract research organization and was fortunate enough to become involved with GLP field studies in the mid to late 1980s for numerous sponsors conducting studies in the United States. This provided the opportunity to experience the field GLP evolutionary process from an industry-wide perspective. Even though practically all sponsors started in the same place interpreting the GLPs for field studies in the late 1980s, it has become quite apparent in 1998 that there is a distinct range among the more and the less overall GLP compliant sponsors throughout industry. In comparison, I have been involved with international multi-country GLP field studies since 1990 but have only worked with one or two sponsors each year since that time so it should be noted that this article is neither from an industry-wide perspective nor is it from any one given year. This chapter addresses problems and situations I experienced with one or two sponsors a year over a period of seven evolving GLP years. Some of the problems from the early 1990s were sponsor specific in nature or have since corrected themselves due to the increase in GLP comprehension and logistical functioning of the various sponsors since that time. The field portions of the GLP studies discussed here were conducted in Central and South America.

The first major consideration for the sponsor or registrant is to determine its submission priorities and to consider all of the various OECD and/or U.S. EPA FIFRA GLP issues involved in conducting these studies. Recently proposed revisions to the OECD GLPs for European submissions as well as the proposed changes to consolidate GLP's applicable under FIFRA and TSCA for U.S. submissions has significantly added complexity to all of the existing interpretation and implementation concerns. Passage of the Food Quality Protection Act (FQPA) in August 1996 will also have a profound affect on sponsors' and registrants' submission priorities.

Many studies in the past have had OECD and U.S. EPA FIFRA GLP regulatory guideline aspects that overlapped and caused confusion in overall study design, coordination, implementation, control, and reporting. Some studies were submissions to various governmental regulatory agencies in Europe, others were U.S. import tolerance submissions while a few sponsors attempted a single study where the submission would suffice for both U.S. and European regulatory guideline requirements. Problems happened primarily because European based sponsor GLP analytical facilities (where study directors and their managements reside) have required their U.S. based agricultural subsidiaries to conduct GLP field studies in Central and South America. Occasionally, the U.S. subsidiaries have had to conduct the analytical portions of the studies but reported to their European based sponsors while their own or South American subsidiary counterparts implemented the field portions of the studies. Coordination among all parties has been inadequate. Whatever the scenario, the study final reports, whether submitted to U.S. or European regulatory agencies generally 1) have not met all their respective OECD or U.S. EPA GLP submission requirements, 2) have been rejected due to contamination, or 3) have had study results that could not be reconstructed nor justified. One study, to meet the U.S. requirements where there was no existing registration, had initiated the field phase prior to the European based analytical laboratory having had properly developed the analytical methods. This particular new petition for tolerance as a U.S. submission had problems once the analytical methods were developed and the field phase neared completion because the methods were not able to be independent validated in a timely manner according to the U.S. EPA requirements. (OPPTS 860.1340 Section C.6.(i).) It is very important that the sponsor determines and researches its submission priorities in advance before study initiation begins.

Once submission priorities have been established, the next major consideration for the sponsor is to determine the various internal players most qualified to adequately perform international study conduct. Study directors, protocols, and study directives many times originate at a European based sponsor location. However, the analytical and field portions of the study can involve the U.S. or Central/South American subsidiaries.

No matter where the field or analytical portions of the study are conducted, it is important that there be direct and timely interaction among the individuals involved in the study. If the sponsor believes that it cannot successfully conduct the entire study in a timely manner, serious consideration should be given to involve contractors. If contractors are involved, the sponsor and contractors should make absolutely sure whose SOPs should take precedence. This should be stated in writing. At present, there is a lack of competition in the number of qualified contractors able to assist sponsors in conducting the field portions of international studies. However, sponsors may have few options other than to involve these contractors if their own internal personnel do not have the expertise, equipment or GLP training to adequately perform the study requirements in a timely manner. Therefore, sponsors should place a high priority on study monitoring and quality assurance to adequately access the performance of all outside contractors whether larger management firms or smaller independent field contract research organizations.

Problems and Solutions

The following are some key study problems with related solutions that are common in conducting international multi-country field studies.

Problem #1 - Lack of Study Director Control. The most significant problem has been a general lack of study director control concerning the field portion of the study. Study directors are usually based in an analytical GLP environment and seldom visit multi-country field sites on another continent due to financial or time constraints. Real time decisions in the field that might affect analytical results are usually left up to the discretion of the field principal investigator. Documentation and reporting of these decisions to the study director in the past were well "after the fact." Many SOP deviations were not properly reported nor acknowledged by the study director.

Solution #1 - Delegate Study Monitors. A solution is to appoint study monitors who regularly report to the study director. Usually QA individuals have limited direct international field experience and are not allowed by GLPs to be involved in any aspect of study conduct. South American field principal investigators are knowledgeable in their respective areas but usually lack sufficient GLP training. A GLP-oriented study monitor with field experience can provide initial GLP training to key field personnel, assist in making real time decisions in the field, and keep the study director informed of major developments that could possibly influence analytical results. Study monitors are usually independently contracted individuals or persons associated with management firms contracted by the sponsor.

Problem #2 - Inadequate Study Plan. Study plans have been inadequate in the past; however, the study plan is the most critical aspect regarding multi-country field studies.

Solution #2 - Spend Money on the Study Plan. Extra money and time spent in preparation during this phase can eliminate more than 75% of the problems before they occur once the study has initiated, as well as provide clear protocol objectives, assignment of responsibilities, and specify documentation procedures. Sponsors can either spend extra money up front during the study planning phase or spend extra money during the later stages of study conduct in resolving problems that otherwise would have never occurred. The significant difference in the latter case is that the quality of the data is usually compromised.

Problem #3 - Unclear Responsibilities. Many field principal investigators were confused as to their proper SOPs, documentation and reporting procedures as well as specifically defined responsibilities such as sample chain-of-custody procedures. Test substance characterization and stability mishaps have occurred because of confusion in delegating European or U.S. based analytical responsibilities.

Solution #3 - Define Duties/Follow SOPs. Study directors, monitors, analytical and field investigators, QAU and support personnel should all be aware of their respective duties and when and to whom they are to directly report to. This should be stated in writing. All parties should operate under the same revision set of SOPs in their own respective languages.

Problem #4 - Lack of Initial Research/Advance Preparation. Lack of advance preparation and research regarding the field crop can lead to misinformation concerning cultural practices and application or sampling intervals.

Solution #4 - Keep Everyone Involved. Everyone should have input during the study plan and protocol development phases. Many European based sponsors are reluctant to have their South American personnel directly involved in study conduct. However, these personnel are crucial in making logistical and financial arrangements with their contacts such as aerial pilots and crop growers or plantations where the test systems are to be located. South American sponsor personnel can assist qualified GLP field contractors in documenting test system pesticide histories and in preventing contamination during the field phase.

Problem #5 - Chain-of-Custody Issue. Chain-of-custody issues are probably the most challenging and problem-oriented areas throughout the entire multi-country field study. Problems that should be noted are sample storage stability questions because of miscommunication on how the samples were to be shipped (frozen or unfrozen), improperly identified or stored samples, problems with South American airlines transporting samples and transporting samples through U.S. customs destined for Europe.

Solution #5 - Maintain Tight Control. A designated individual should be assigned to thoroughly research and prepare the necessary documentation concerning the entire chain-of-custody process. Precise instructions and responsibilities should be given to the personnel handling the samples during the field phase. If necessary, someone should escort the samples through customs to prevent unnecessary delays.

Communication is critical when conducting international multi-country field studies. The following are some of the communication problems with related solutions. Lack of communication has caused confusion and logistical mishaps in study conduct from analytical laboratory locations to remote field test system locations in South America.

Problem #6 - Unclear Study Directives. Unclear study directives have occurred when there was not enough advance preparation by internal study management and research staff. In one case, study directorship responsibilities were transferred by the sponsor from Europe to the United States because of work overload. Late protocol amendments were issued concerning changes in the sampling requirements confusing field personnel. Samples had already been collected in some countries but not others.

Solution #6 - Design the Study Plan Together. The best solution is to avoid the problem and devote as much time as possible in defining and assigning study directives during the study planning phase. This should include planning meetings with key field-oriented personnel. Document what is being communicated.

Problem #7 - Language Barriers. Having three or more languages involved in multi-country field studies is not uncommon.

Solution #7 - Multi-Language Protocol. Protocols, amendments, and SOPs should be in multiple languages. The official protocol (original signed by study director) should be accurately translated and include all applicable languages. Deviations generated from the field locations should be written by the individual most closely related to the event (usually in Spanish) and accurately translated prior to being submitted and signed by the study director. Translators should be available if necessary to assist with the field portions of the studies.

Problem #8 - Vague Protocol Instructions. Many protocols are not precise enough to provide clear directives for the field phase. Documentation and reporting procedures need to be explicitly defined.

Solution #8 - Define Protocol Terminology. Protocols are evolving to now define terminology of key personnel such as principal investigator and their responsibilities. This is necessary to state by name the individuals responsible for applicator and supervisory duties over several country locations simultaneously conducting applications and samplings. Protocols also specify QA reporting procedures, describing time frames and designating distribution lists. Deputy Study Directors were at one time part of this distribution, but European based sponsors no longer have these job description positions which caused confusion during QA reporting. Protocols can be more than 50 pages, include sample lists, and describe by name detailed information, such as, the exact tools to be used or type of gloves to be worn during field sampling.

Problem #9 - Insufficient Training. Field personnel seldom are supplied with the proper equipment or training prior to study initiation. Communication problems are enhanced by those individuals who are not familiar with GLP research and are shy in conveying their concerns and needs due to language barriers.

Solution #9 - Advance GLP Preparation. GLP studies are expensive; however, it is very cost effective to properly supply and train field personnel. This increases the confidence of those individuals to perform their assigned duties and significantly reduces the possibility of sample contamination.

Problem #10 - Lack of Documentation. Field generated data have been difficult to reconstruct or verify. Confusion in proper rate calculations and recording of actual measurements was apparent because European based sponsor protocols were written using metric units while South American field personnel were required to convert to English units because they were supplied with field trial notebooks by the U.S. subsidiary.

Solution #10 - Study Specific Notebooks. Study specific notebooks prevent language and rate/unit discrepancies. The notebooks should be reviewed in advance by field study personnel and, if necessary, revised to include detailed information concerning application and sampling critical events.

Problem #11 - Inadequate Reporting. Many of the critical phases for test system locations in South America were not properly quality assured nor monitored.

Solution #11 - Increase QA/Study Monitoring. Quality assurance and monitoring of the first application event in each country should be required. Some locations may have more than ten applications each so proper monitoring and reporting should begin with early applications as well as the sampling critical phase.

Conclusions

When I first became involved with international multi-country field studies, I contacted several individuals at EPA to obtain advice concerning the regulatory audit aspect of the field portions of the studies. EPA officials informed me that any GLP study submitted to the Agency is subject to a regulatory audit under the same inspection procedures no matter if the field locations were within or outside the continental United States. The importance was stressed of having qualified and trained personnel, updated SOPs and a detailed chain-of-custody for the test substance and samples. Agency officials acknowledged that the "infra-structure and general GLP awareness" was not the same in Central or South America as in the United States, but that the field phase should be conducted "in principle" according to the GLPs.

Since 1990, I also have had the opportunity to meet and discuss the conduct of these studies with European regulatory officials. They also reiterated that the GLP submissions to various European regulatory agencies are subject to detailed regulatory inspections.

As for the crop protection industry, registrants and sponsors must continue to interpret and implement the OECD and U.S. EPA GLP challenges by re-examining their product development and product defense strategies while generating sound, scientifically defensible data.

Therefore, international multi-country field studies conducted under GLPs need to be properly implemented and documented. In order to achieve this, the sponsors need to 1) concentrate their efforts on the study plan and delegating responsibilities, 2) provide a clear, detailed multi-language protocol, 3) insure proper training, communication and documentation procedures, 4) provide qualified quality assurance and study monitoring, and 5) involve contractors if deemed necessary.

Chapter 22

Field Trials in Latin America: Se Habla GLP?

Steve West

Research Designed for Agriculture, 2246 West 19th Place, Yuma, AZ 85365

Agricultural research work that is conducted to support pesticide import tolerances in the United States and the European Community must be done with increasing regularity in Latin America. Performing such research in full compliance with the FIFRA and OECD Good Laboratory Practice requirements poses unique challenges in Latin America. Presented here are some of the obstacles and challenges one might face when conducting this type of field research.

As the pesticide and food safety regulatory systems of the United States and Europe have matured, the issue of residues in imported food has taken a higher priority. Non-Latin countries now regularly require field residue trials to be conducted in Latin America on commodities such as bananas and coffee. The United States EPA's policy of requiring 12 GLP field trials for bananas, or Europe's benchmark of 8 GLP field trials for "major crops", such as bananas, has added a new wrinkle to getting the work accomplished.

Without a doubt, a network of field researchers, both within the major chemical companies and the small contract research companies, does not exist in Latin America to the extent that it does in Europe or the United States. In Latin America, traditionally, the government agencies, such as the various universities or the federal or state run agricultural research services, have conducted virtually all of the field research except those projects conducted by sponsor staff. With the move of importing countries to require GLP field work in Latin America, there is a lack of infrastructure to accommodate the work required. Unfortunately, the reality is that the agencies, such as the universities and state run research services, are seldom in a situation to successfully conduct GLP studies. Sponsors, by and large, had quit using public agencies for GLP studies years ago, primarily, because the mission of these agencies and the expertise available were not consistent with the extensive requirements of GLP.

This has led to the necessity of sponsors either to go to the arduous and time-consuming task of training their own people or hiring out their work. For companies that have large, on-going programs, setting up an in-house GLP unit is an option. But in Latin America, as in the United States and Europe, the major companies are discovering that maintaining a current, effective in-house GLP field staff is very expensive. The reality is that unless you do GLP work all the time, you fall out of practice. The extremely high levels of compliance required by the sponsor Quality Assurance Units and by the regulating agencies makes documentation mistakes unacceptable. The learning curve is now over, and the work must be either near perfect or it is rubbish.

Over the past 7 years, Research Designed for Agriculture has been conducting GLP field trials in various Latin countries. Our experience has led us away from sub-contracting fieldwork and now RDA conducts studies with its own, multi-national staff based in Yuma, Arizona. This paper attempts to convey some of the challenges and obstacles of accomplishing GLP fieldwork far from home.

Components of Conducting a GLP Field Trial in Latin America

What is GLP? For the purposes of field research, GLP is really very simple. The trial must be conducted according to a protocol, using SOPs, and with ample documentation to be able to reconstruct everything that happened. This is critical because unforeseen things will happen.

Reliable Information. The biggest single obstacle in getting started with field trials in a remote, or "off site" area is getting accurate information. This is the case whether you are doing a field trial across the county or across the ocean. The difference is that when you are across the ocean, poor information has the potential to be much more disastrous. The consequences of flying to another continent in anticipation of finding flowering grapes, for example, and then finding when you get there that they flowered 10 days ago, can be catastrophic. Obtaining reliable information on everything you can imagine is, without a doubt, the most important, the most critical, and yet the most difficult component to conducting a field research trial.

Why is this? It would seem that this should not be so difficult. The problem lies in the foundation of viewpoint. Most often the people we contact for information are not other small plot researchers. Often contacts are in company headquarters or tied to a desk. Even those who are out in the field are usually traveling to many areas and viewing many crops. Their information may be well founded from a field or area, but it is not accurate for the location where your plots need to be located.

Many times the growth stages or timings are somewhat ambiguous to those not accustomed to GLP residue trials. Frequently, the information we are asking for is not something people normally think about. For example, in most areas the broccoli goes from seed to harvest in about 120 days, but ask someone, even someone very familiar with broccoli, exactly what the growth stage will be six weeks before harvest, and you will get a range of answers. Since the harvest window on broccoli is only 10 days or less, being off two weeks is a significant problem.

Protocol. Writing a proper protocol for the conduct of an "international" field residue study is somewhat more difficult than for a domestic study. Most commonly we work with studies that are being run by either a United States company or by the United States branch of a company. The protocols and SOPs referenced are all written for trials executed in the United States. They don't take into account the lack of infrastructure in many foreign countries or the scarcity of many items and services commonly found in the United States. These differences account for at least half of the GLP problems on international studies. Marcus Jensen discusses the topic of protocols and related GLP issues in greater detail in the next chapter of this publication.

Site Selection. With some studies we are allowed the option of choosing any location in one of several countries; however, in other studies we must stick to areas specifically required by a regulatory agency.

Site selection is broken down into two basic decisions. The first decision is the general area. For example, if we are doing tomato trials in Mexico, we need to decide if the trials should be in Sinaloa, Baja California, or Michoacan. With banana trials, the choices are broader. We decide if they are to be done in various countries, and then within the countries, we look at the various areas.

Personal Safety. Unfortunately, personal safety has become a significant issue for site selection in some areas. Whenever a new project is discussed, the question of personal safety moves to the forefront. At RDA, we have a policy of not taking projects in areas we perceive as dangerous. Columbia, parts of Mexico, and Bosnia, among others, make this list. The wisdom of this decision was reinforced in early 1997 when an American was kidnapped while conducting field trial GLP quality assurance inspections in Columbia. (After paying the ransom, he returned home.) Areas with high incidence of cholera, malaria, etc., are also avoided.

Field Considerations

After the general area is determined, a decision on the exact field location must be defined. Sometimes this is a field that is custom grown for the experiment, other times the work is conducted in a commercial field. In either case, finding someone you are confident will work with you today and tomorrow is critical. However, it is not just enough to feel good about the manager or owner. The worker making 5 dollars a day in the field has just as much (or more) influence on the outcome of your trial as the owner. It is important to talk to and get on good terms with the foreman and workers in the sections of the farm you are working.

Field Practices and History. Identifying a site with the proper field history can sometimes be challenging. In most Latin countries, the economies are such that price is the over riding factor in pesticide usage. Consequently, many of the older, off patent, organophosphate and carbamate insecticides are heavily used, as well as the lower priced fungicides such as EBDCs. If a protocol restricts either the history or current season use of one of these products, you may well be forced into the expensive choice of having a field custom grown. It is important that the protocol and appropriate SOPs, which reference the history, are referenced and appropriate GLP documentation is provided.

Current Season Practices. These are equally important to know. If you are planning a corn study that will be harvested for grain, be sure that the grower doesn't cut at normal silage time, and that alfalfa for that type of trial will be cut and allowed to dry for hay and not all cut and sold as fresh forage as in central Mexico.

Support Services. If you will be using support services such as aerial applicators, commercial freezers, trucking companies, etc., be sure you talk with them prior to starting. Having an aerial applicator that will work you into the schedule and treats you as a client is valuable. Working with an applicator who puts you off as a nuisance is often fatal.

Access. Access is an underestimated challenge in international work, and it means more than good roads to the field site. Do you have access to hotels that aren't always overbooked? Have you found access to places that supply CO_2 for your sprayer or dry ice to freeze samples? How about rental car availability? This already assumes that you can fly into a reasonably close location, and that the one flight that day was on time, or even running. In a two-month period one fall, we had to delay starting trials by a week because of rained out roads and needed to scramble several times because the cars we reserved (and "guaranteed" with Hertz) were not available when we arrived.

Access problems of all sorts will cause delays and challenges that are seldom foreseen. In Costa Rica, on three different occasions the main road from San Jose to Limon was closed due to land slides on the nights I was trying to meet an early morning spray appointment. Is driving and extra 3 or 4 hours into the wee hours of the morning so you can sleep for 2 hours before meeting the pilot at 5:30 a.m. what you had in mind? Will the sponsor be receptive to the excuse that the application was not on schedule, for the third time?

Trial Layout

Trial layout brings on a set of unexpected challenges. Finding a suitable permanent marker can be frustrating and often takes some imagination. In many areas, the high amount of pedestrian traffic will thin out the plot flags. We have had some plots that have required reflagging every time we came to the field. On one occasion, we were marking off a 2000-meter long plot in a banana field by carrying a 100-meter tape. The person in front would stick in a flag at the end of the first 100 meters and the person following would move up and stop at the flag, then the person in front would move on and place another flag at the end of the second 100 meters. As soon as we were finished, we walked back down the farm road to collect the flags, which were already gone!!

When it comes time to put in area maps for general reference, you may find that none exist. I have been places where I think that everyone in the area was born with an innate road map. Everyone knows where even the most obscure place is, but no one has ever even seen a map of the area, let alone know where to buy one. I recently had to write in a trial notebook "No local map available, and this is one of those places you'll have to be taken too, as you can't really get there from here without a guide".

Without proper GLP documentation in the trial data, solving some of these challenges would be impossible. Actual maps of how to find the plots on the farm are critical. Care should also be taken to be sure that sponsor required items in the protocol and required SOPs are identified and addressed now to avoid mountains of problems later.

Test Substance

Bringing experimental compounds into a foreign country can be challenging. Bringing in small quantities of GLP characterized lots of a labeled compound into a foreign country can be just as big of a challenge.

The source of the shipment is significant. Shipments coming from the United States, if the product is not labeled, have to carry the warning that the product is not approved in the United States, which makes many countries nervous. In other cases, the sponsor company's import permit for a labeled product is for product coming from France, for example, but the GLP characterized product comes from the pilot plant in England. This creates more logistical hoops to jump through.

Another aspect of small lots it that they are labeled differently. Even if there are permits in place for either the commercial or experimental product to be imported, usually the people filling out the paperwork for the permit are not the ones in the lab labeling the containers. The containers of Test Substance arrive, labeled with the essential GLP information, but it is not the same information. The import permit says "methyl acidiazole", but the container is labeled "RD-4489".

In some countries, the delays in customs can be interminable. On one occasion in Costa Rica we had to wait 30 days for the sponsor to get the product cleared from customs. We had no idea what the storage conditions were in the customs house, and as is normally the case, the sponsor did not include a thermometer in the shipment. The sponsor was sure that this oil suspension product was stable, so we pulled a sample for analysis and started the trial. After the analysis came back, we did the trial again with a new lot.

It also is important to use the correct formulation. Too often we do trials with a substandard formulation. Starting an international trial with an experimental formulation is foolishness. On the airstrip in Guatemala is not where you want to find out that your new formulation settles into a concrete cake at the bottom of the 200-liter drum provided for the air trials we are doing.

Once you get the correct formulation there, you have to find a place to store it. It needs to be secure and accessible. Often the sponsor's people want to keep it in their warehouse or at one of their distributor's warehouses. However, they may not be around at 5 a.m. Sunday morning when you need it or be available Saturday night at 10 p.m. after you drove back in from the field and got delayed by an auto accident on the highway? What do you do with it now? Your flight is at 6 a.m. the next morning and everyone you know is out of touch!

The question of temperature during storage is fairly straightforward as recording thermometers or min/max thermometers can be used. Finding a temperature-controlled facility is not realistic. If you have a product that can't be stored over 90 degrees; you may be in trouble. Warehouses in the desert areas of Chile will regularly be 120° F in the afternoons. Unless the product containers are small enough to be kept in a refrigerator, keeping the Test Substance cool under such conditions is next to impossible. Even if you keep them in a refrigerator, power in many of these areas is subject to frequent outages. If the refrigerator is out on the porch in the sun and the power dies for two days, you will have a high peak temperature. If you are using a min/max, and are there only once every two weeks to make an application, you have a lot of explaining to do, and possible re-analysis of the compound.

Once the trial is completed what to do with the left over Test Substance? Better yet, if it is a United States EPA trial, what do you do with the containers?? Re-importing them into the United States is difficult unless they are really clean. If you have any significant amount of Test Substance left, disposal is challenging. There are no easy answers to this, and every situation is different, but the issues must be dealt with.

Application

The application equipment to be used is driven by the protocol, crop and plot size. Air trials in bananas require aircraft, and airblast applications to citrus require turbine blowers. Post harvest applications almost always require a unique set up, anything from a 5 gallon bucket dip tank to a commercial 300 PSI micro mist chamber. Field crops are usually treated with a conventional boom sprayer. If the plots are to be large, generally, commercial equipment will be used. If the plots are small, then small plot equipment is called for.

Our experience with commercial aircraft in Latin countries has been good. As an example, this past year we have been working off a runway in Costa Rica where I counted 8 turbine Thrushes one morning. Added to the three radial Ag Cats, I calculated there to be about 5 million dollars worth of aircraft there. These aircraft are state of the art, and the pilots operating them are first rate. The biggest hurdle to cross is getting an applicator that is available when you need them and willing to put up with the idiosyncrasies of GLP work. Finding an aerial applicator willing to do this every two weeks, even when they are busy and behind from five days of rain that has kept them grounded, is truly challenging.

When field trials require airblast or large plots, you're normally forced into borrowing equipment. Our experience with borrowing grower equipment has been less positive. Tractors without any tachometer make repeating speeds and spray pressure (on PTO driven sprayers) an art form. Functioning pressure gauges are on the endangered species list. You can count on pumps leaking, or being worn out, and the spray nozzles are seldom the same size, let alone in new condition. It is very important that you know the equipment intimately and are prepared to completely overhaul it if you want uniform spray patterns. We used a grower's sprayer in Chihuahua in 1996 where we brought down 300 feet of hose, 30 new pressure relief nozzle bodies and all new spray tips. This $500 investment was well worth it as the grower liked having his spray system completely redone, and the amount of time we saved in calibration and fixing leaky hoses and nozzles was immeasurable. Not all grower equipment is in poor repair. We have found some cooperators where the equipment was in top shape; however, that is the exception, not the rule.

In other cases, we are supposed to mimic local practice for application methods. Since labor is inexpensive in Latin America and equipment is not, the spray methods follow suit. Usage of the "Solo" type hand pump, single nozzle backpack sprayer is very common. In Central Mexico, the technique for spraying cabbage is to spray each head until runoff, with even more run-off on the plants damaged by worms. How do we do a uniform GLP application like that? Pole tomatoes are similarly handled, with crews with backpacks waving the wands up and down and pumping up the pressure every few meters as they walk the 1 meter tall rows. These methods may be effective controlling the pests, but when you are measuring parts per billion of residue on the fruit, you want to be more uniform in your methods.

Whenever possible, we try to use our own equipment for plot spraying. It is designed for the purpose, and we know the condition of it. While it is sometimes possible to borrow small plot sprayers (normally CO2 backpack sprayers), we try to avoid that since there are seldom logs telling you what was in the sprayer last or what type of cleaning is needed to clean it up.

Mixing the chemical often presents problems. If the product is a dry powder, how will you weigh it? What if the balance you brought dies? What will you be using for standard weights? If the product is a liquid, things are easier, assuming you use plastic cylinders or syringes. If you like using glass cylinders, have in mind a good source where you can get replacements on Good Friday when yours is in pieces. Many protocols are now asking for potable mixing water. More than once I have stopped and bought a 5 gallon "water cooler" bottle to take to the field since only pond water was available there. When slurries need to be made, or water of a certain temperature is to be used, additional equipment is needed. Making small batches for big machines like aircraft also requires some forethought. We regularly use 200-liter drums with the lid cut off for small mix tanks and for agitation we have several electric trolling motors for boats in stock. The big mixers found at airstrips often won't even prime with our mixes.

When you do bring your own equipment, a safe place to leave it where you can get to it when you need it is required. The same comments about test substance availability apply here as well. Another issue to deal with is whether you need the sprayer elsewhere. Not many of us can afford to have several sets of $1,500 sprayers and related equipment scattered all over Central America at the same time. If you need to move it around, getting CO2 is an issue, as is paying the fees and obtaining the permits to import it into the country. We spend hundreds dollars a year on import permits for Mexico alone, just for our equipment.

Invariably, when you move equipment around, and even if you don't, it will eventually break. Having the tools you need, and the parts required is critical. Finding a replacement part for a specialized sprayer in Honduras may be impossible, and if not, it will consume time at the least. Carrying pipe threaders, nozzle bodies, gaskets, gauges, etc., etc., is not a luxury, it is essential.

Making the application is about the same as anywhere, except that you are a long way from home and on a tight schedule. Windy days, hurricanes, rain storms, wet fields, labor crews in the way, etc., are all problems to deal with. The hitch is that when you get delayed you get to rebook your flights, hotels, rental cars, etc. It becomes expensive. If you have other trials you are moving to after the one your working on now is done, you may incur ill will for the people waiting for you there, and Sponsor A may not think that waiting out a storm for a trial with Sponsor B is a good reason to deviate on their schedule.

Sampling

Unquestionably the toughest part of sampling is having the right crop stage to sample. With one of our first trials with Cauliflower in Arizona, we were working in a commercial field of Tanamura and Antle, the second largest vegetable company in the United States behind Dole. We wanted to set up our spray schedule based on their predicted harvest maturity, after all, they are the experts. We asked, "When will the crop harvest be so we can start our sprays 6 weeks before then?" The answer? "Your guess is as good as ours, we missed the harvest dates by 30 days last year!! The weather turned cold, and we were off!". Last year in Arizona, the early head lettuce took only 65 days rather than the usual 80 because it was hot. This unpredictability plays havoc on long distance trials especially, because you are normally not there all the time to see the delays or speeding up of the development of the crop.

Assuming you are lucky enough to be close on your timings and the crop is ready when the trial is, getting dry ice to freeze or ship samples, freezers to store retain samples in (essential), finding the labor to assist you when needed, arranging for the combine or other commercial harvest equipment you need, and so on requires patience, forethought and often, luck.

After the samples are in the ice chests and ready to go, you either need to have a good freight forwarder that can assist you in shipping the samples to the destination country, or you need to carry the samples as baggage. We normally prefer to follow the baggage procedure, as we have the permits needed to import frozen Ag products, and it is much quicker to hand carry them through Customs and Plant Quarantine inspections. The catch is that you are limited on the amount of dry ice you can carry on a plane, so you'll need to have the samples super frozen so they can remain frozen with the approved quantities.

Having a good supply of dry ice available is really helpful. Over the years we have located virtually all of the dry ice suppliers in Latin America. Even with good plans, samples get lost, so it is quite important that you have a reserve set of samples reliably stored frozen where you can come get them in the event of a problem later.

The cost of sample shipping is very expensive. It is not unusual to spend five thousand dollars on shipping samples over seas. We recently spent over one thousand dollars on excess baggage for hand carrying some samples. It would have been twice that, but the Mexicana people gave us a break since we flew out of that airport so much and they knew us.

Data Documentation

In order for any of this work to be acceptable, it needs to be GLP. This is much easier if your SOPs are written flexibly enough that you don't have countless deviations in the normal course of work. Normally, this means not using the sponsor SOPs. You will need to make up training logs on the spot for pilots, harvesters and anyone else who helps you in the field. Developing equipment logs and documentation in the data of all the various equipment such as freezers and sprayers that you borrowed is a given.

As in the United States, plan on all field data except that which you generate to be nonGLP. Sometimes the field history data we can obtain in Latin America is better than the same data obtained in the United States. Commonly though, fields are rented, so you often have a situation where the last farmer for that field is unknown. If the sponsor has a problem with this, the time to know it is prior to the work being conducted. Unless there is a potential chemical conflict found from the use history, it is a noncritical requirement; however, it is amazing how some QA people, in particular, believe the world is ending if we don't know how many times an insecticide was sprayed on an onion crop three years ago.

Availability of weather data, GLP or otherwise, can be challenging. Mexico has an excellent weather data collection system, on a par with NOAA in the United States. It took us a couple of years to find the right people to call, but now that we have that connection many problems are solved. You may have to settle for weather that is a long way away, however, especially in countries other than Mexico and Chile. The concept of on site data is great, especially if you have a big budget for replacing expensive automated equipment. When you cannot keep plot flags in the field, there is little chance of keeping a $1,200 automated rain collector. Now that EPA is just asking for a general statement regarding the weather compared to "Normal" there is less need for on-site data.

Conclusion

Conducting a good quality GLP study in Latin America is possible if you are willing to make the commitment to make it happen. To do so successfully takes a more flexible mind set from the study director, extensive planning, qualified people who can make something out of nothing, imagination to make do with what is available, and, of course, luck. No amount of planning and money can buy a sunny day. The realities of culture, distance, and nature validate the agriculture version of Murphy's Law "Some things will go wrong, we are just never sure which things." and then the International Anecdote "Murphy was a damned optimist".

Chapter 23

The Canadian Pesticide Registration System in the Context of International Harmonization

Daniel Chaput

Pest Management Regulatory Agency, Health Canada, Ottawa, Ontario K1A 0C6, Canada

The Pest Management Regulatory Agency (PMRA) of Health Canada is involved in a wide range of international harmonization initiatives both under the North American Free Trade Agreement (NAFTA) and under the Organization for Economic Cooperation and Development (OECD) Pesticides Program. Some of these initiatives, such as, joint reviews, development of common data requirements, and electronic data submission, are briefly described. The development of joint residue zone maps covering both Canada and the U.S. and the implementation of OECD GLP are discussed in greater detail.

The Pest Management Regulatory Agency (PMRA) was established in April 1995 in response to the recommendations of the Pesticide Registration Review (PRR) Team. The Multistakeholder Review Team was charged with studying and making recommendations to improve the federal pesticide regulatory system. Administration of the Pest Control Products (PCP) Act was transferred from the Minister of Agriculture and Agri-Food to the Minister of Health, while pest management regulation resources and responsibilities from four government departments were consolidated in the PMRA.

PMRA has responsibility for protecting human health and the environment while supporting the competitiveness of agriculture, forestry, other resource sectors and manufacturing. The Agency is also dedicated to integrating the principles of sustainability into Canada's pest management regulatory regime.

Recognizing the clear benefits of harmonization, the PMRA is currently participating in harmonization activities which support the development and implementation of common data requirements, common test guidelines, common approaches to risk assessment, as well as common review formats and the format for regulatory decision documents.

At present, the PMRA is pursuing a number of bilateral and trilateral initiatives among the three North American Free Trade Agreement (NAFTA) countries and with a range of other countries through the OECD Pesticides Programme. Some of the initiatives underway within these two forums are summarized in this paper. More details on PMRA harmonization activities are available on the PMRA internet site (http://www.hc-sc.gc.ca/pmra).

While certain harmonization projects are specific to NAFTA, others are pursued concurrently in both the NAFTA and the OECD forum. In some cases, a project initiated under NAFTA could later move under the broader OECD umbrella.

Harmonization Activities under NAFTA

The NAFTA Technical Working Group (TWG) on pesticides is the forum for harmonization activities between regulatory agencies in Canada, the United States and Mexico. The goals of the NAFTA TWG are to: 1) share the work of pesticide regulation; 2) harmonize scientific and policy considerations for pesticide regulation; and 3) reduce trade barriers. Specific activities carried out under NAFTA include:

Development of Harmonized Data Requirements. The harmonization of data requirements is one of the cornerstones for efficient joint reviews and worksharing of pesticide evaluations. Data requirements have been harmonized in some areas while significant progress is being achieved in others. Some of the areas in which data requirement harmonization projects are being conducted in consultation with the United States Environmental Protection Agency (EPA) and, in some cases, with other countries include:

Product Chemistry Data Requirements. The PMRA published in July 1997 a set of revised product chemistry Regulatory Proposals (*1,2*) describing the data required to comprehensively characterize technical grade active ingredients and end-use products. These revised documents are harmonized with the US requirements (EPA Product Properties Test Guidelines Series 830).

Residue Chemistry Data Requirements. The PMRA published in June 1997 a set of residue chemistry guidelines (*3*) describing the data required to evaluate and assess the nature and magnitude of the residues in foods, perform dietary exposure assessment and establish maximum residue limits. Also included in these guidelines is specific guidance related to supervised crop residue trials conducted in Canada and/or the U.S. The guidelines describe the Canada/USA residue maps which delineate regions that are unique and which are consistent between the two countries, allowing for the data produced in equivalent zones in either country to be used in support of registration. The Canada/U.S. residue maps are discussed in more detail in the last section of this paper.

Data Requirements for the Registration of Microbial Pest Control Agents and Products. Substantial progress has been achieved on the harmonization of data requirements for these types of products especially in the areas of characterization and health risk assessment. Further harmonization is currently underway through the OECD pesticides programme. A PMRA Regulatory Proposal describing these requirements is scheduled for publication in late 1998.

Joint Reviews of Pesticide Data Submissions between Canada and the U.S. A key component in supporting the various harmonization activities is the practical experience obtained through an expanding programme of joint reviews between Canada and the U.S. and work sharing with a broader range of countries. This provides invaluable experience in refining the terminology, the level of detail, and really understanding the significance of any apparent differences in the interpretation of data.

The PMRA and the EPA have established a process for the Joint Review of chemical products that meet the EPA criteria of "reduced-risk" and products containing either microbials or semiochemicals (including pheromones). Candidates eligible for Joint Review must contain a new active ingredient with use patterns common to both countries and be supported by complete data bases. During the course of a Joint Review, the PMRA and EPA together may request additional information or data.

It is anticipated that the Joint Review programme will increase the efficiency of the registration process, increase access to pest management tools in both countries and facilitate the registration of alternative pest control products with a reduction in review time but not in safety standards. The targeted time frame for the evaluation of a Joint Review is in the order of twelve months as opposed to eighteen months for a standard submission.

Joint reviews of a chemical, a pheromone and a microbial pesticide have been initiated. Since this Joint Review initiative is a first for both Agencies and for industry, a number of adjustments are being made as it proceeds in order to streamline the process to the extent possible.

Harmonization Activities under the OECD Pesticides Programme

The OECD's Pesticide Programme is one of ten subprogrammes of the OECD's Environmental Health and Safety (EHS) Programme. The purpose of the EHS Programme is to help countries manage the risks of chemicals as efficiently and effectively as possible while enjoying the many benefits they provide.

The PMRA represents Canadian interests in the OECD Pesticide Programme. Canada has taken, and continues to take, an active role in many projects within the OECD Pesticide Programme. These include:

Harmonization of Guidance Documents for Industry Data Submissions (Dossiers) and Country Data Review Reports (Monographs). The goal of this project is twofold: i) Through harmonized data requirements, to develop a common data submission (dossier) acceptable to all OECD countries; and ii) To develop a common format and content for individual study reviews and final Monographs prepared by countries.

The preparation of a single dossier acceptable to OECD countries will increase the efficiency of national regulatory processes by facilitating work sharing among countries and could result in substantial savings to industry. In addition, information exchange among countries will be facilitated with the adoption of common review formats. A meeting among representatives of national governments and industry was held in January 1997 to initiate discussion on the development of an OECD document based on that prepared for use in the European Union (EU).

Along the lines of common data submission and as a result of the OECD forum activities, the PMRA introduced in 1996 a requirement for the provision of comprehensive data summaries in registration submissions involving major new uses and new active ingredients. The PMRA is using the approach adopted by the EU while at the same time working with the OECD to make the EU format acceptable to all OECD countries. This requirement for a summary is intended to speed up the review process by providing decision makers in the PMRA with a clear, comprehensive summary of the characteristics of the product, its risks and value.

Electronic Data Submission. Canada, through the PMRA, has taken the lead in forming an international group to coordinate a harmonized approach for an electronic capability. The formation of the Global Regulatory Information Technology (GRIT) group was announced at the November 4, 1996, OECD Pesticide Forum meeting. Participants include the pesticide Agencies from the United States, Canada, Germany, United Kingdom, Australia, European Union member states, plus representation from the OECD Pesticide Forum secretariat, the European Crop Protection Association, the American Crop Protection Association and Canadian industry. The development of a compatible electronic submission and review process will increase efficiencies throughout the regulatory process. It is a logical next step from the work to develop common submission formats and data review reports. GRIT will consider developments and will make strategic plans for electronic data submission while ensuring good cooperation at the international level.

Interpretation of Data. A more consistent approach to the interpretation of data between countries is paramount to allow reviews and work to be shared among all jurisdictions. To this end, there are several projects related to the development of guidance on the interpretation of study results including:

- Repeat Dose Toxicity Data Review Guidance (US/Canadian/Australian project)
- Guidance for Summarizing Results from Field Dissipation Studies (UK proposal)
- Aquatic Dissipation Data Review Guidance (Canadian/US proposal)

Good Laboratory Practices (GLP). The PMRA, in line with its goal to pursue international cooperation, is committed to implement the OECD GLP Principles for pest control products in a timely manner. PMRA activities in this area are discussed in more details in the next section.

The Introduction of OECD Principles of Good Laboratory Practices (GLP)

Good Laboratory Practice (GLP) is defined by the OECD as a quality system concerned with the organizational process and the conditions under which non-clinical health and environmental safety studies are planned, performed, monitored, recorded, archived and reported (*4*). The purpose of GLP is to promote the development of quality test data which forms the basis for the mutual acceptance of data among countries.

To date, Canada has no formal GLP regulatory program in place despite some attempts to introduce GLP requirements in the area of drugs several years ago. In 1988, the promulgation of the Canadian Environmental Protection Act (CEPA) created the basis for a GLP program for industrial chemical New Substance Notification Regulations (NSNR), but it has not historically been enforced. An informal voluntary compliance monitoring program is currently in place in this area pending the development of GLP regulations.

The PMRA, in support of its international harmonization initiatives, intends to implement GLP for pest control products in a timely manner. To this end, the Agency published in 1996 a GLP consultation paper (*5*) outlining a proposed approach for establishing GLP requirements for pest control products and for monitoring GLP compliance. The emerging PMRA GLP program is being developed after careful consideration of the comments received on the 1996 proposal.

The Canadian Context. The preamble of the OECD document "Revised Guide for Compliance Monitoring Procedures for Good Laboratory Practices" (*6*) states that *"Member countries will adopt GLP principles and establish compliance monitoring procedures according to national legal and administrative practices..."* Thus, it would appear evident that the OECD recognizes and accepts that there can be some degree of variability in the application of GLP depending on the national context. This conclusion is supported by the variety of approaches to GLP implementation and monitoring currently in place in OECD member countries.

Some of the key elements of the Canadian context which influenced the PMRA approach to GLP include:

i) Although GLP is not formally implemented in Canada, it is already an industry standard given the structure of the pesticide industry originating mostly from the US and / or Europe where GLP has been in place for many years;

ii) GLP laboratory capacity currently exists in Canada as a number of test facilities have been submitting pesticide GLP compliant studies to EPA for a number of years. However, in the absence of a Canadian GLP compliance monitoring program and given the scope of the EPA program, this claim has not been systematically verified by an independent body;

iii) The Canadian government, as many others world-wide, is in the "right-sizing" mode, redefining priorities and encouraging alternative program delivery approaches in light of reduced resource levels; and

iv) The PMRA is currently implementing a cost recovery regime.

Overview of the Emerging Canadian GLP Program for Pesticides. The requirement for all health and safety studies submitted to the PMRA to be GLP compliant will be established through the publication of a Regulatory Directive; a tool used by the PMRA to provide more details and/or interpret the Pest Control Products Act and Regulations governing the regulation of pesticides in Canada. As per the approach used in certain European countries, the requirement for GLP compliant studies will be introduced gradually, by study types, to help ensure an orderly implementation. For example, the principles will initially be applied to crop residue studies (including field trials) in support of the U.S./Canada joint residue zone maps which are discussed in the next section of this paper.

A GLP compliance monitoring authority (CMA) will be established as per OECD requirements. The CMA will be responsible for the administration of a GLP compliance programme and for discharging other related functions such as publishing documents detailing programme operation, ensuring that an adequate number of trained inspectors is available and maintaining records of GLP compliance status of inspected test facilities.

Steps are being taken to recognize the Standards Council of Canada (SCC) as a domestic compliance monitoring authority (CMA) for pest control products. The SCC Conformity Assessment Division currently administers a national laboratory accreditation program which includes over 200 laboratories accredited to ISO/IEC Guide 25. It is likely that a number of these labs would also be involved in the GLP programme for pesticides. In these cases, efforts will be made to streamline the assessment activities in light of the commonalities between OECD GLP Principles and the ISO Guide 25 accreditation while preserving program integrity and identity.

In order to avoid duplication and minimize operational cost, the compliance monitoring program will build on the existing SCC administrative infrastructue to the extent possible. The technical expertise required to carry out inspections will be drawn from staff in government agencies as is the current practice with other SCC programs. The programme will be based upon OECD GLP requirements.

A GLP compliance inspection would involve an on-site inspection in conjunction with an audit of one or more ongoing or completed studies. An inspection report would be prepared and forwarded to a SCC review panel for a decision on certification. The SCC would issue a certificate once the certification is approved by the panel. This formal recognition of a test facility GLP status would address international requirements and broaden market acceptance of the laboratory and their data.

This approach was recognized in a 1994 OECD GLP Panel statement (*7*) which acknowledged that such quasi-accreditation programs are valid if based upon OECD GLP Principles, as opposed to ISO Guide 25, and include government oversight. A similar approach is currently in place in Australia, New Zealand, Ireland, Norway and Sweden, where the National Accreditation body (NAB) is also the compliance monitoring authority. In other countries the NAB is involved in GLP compliance monitoring by acting as a delivery arm for the compliance monitoring authority (e.g.,

France) or an agreement exists between the NAB and the GLP CMA concerning the mutual acceptance of common element inspection data (e.g., Belgium).

After the initial implementation phase, the program will gradually become self-sustaining and function on a fee for service basis between the test facilities and the SCC as is the practice with the Council's current programmes.

Multiple Quality Assurance Systems in Analytical Chemistry Laboratories - A Viewpoint

Quality assurance is certainly not a new concept in analytical chemistry. However, the increasing demand for legally enforceable and decision-oriented data and the need for international acceptability of this information has led to the formalization of quality assurance systems. These systems are codified in guides or standards and include, ISO Guide 25 (EN 45001 in EU countries), Good Laboratory Practices (GLP) and the ISO 9000 series of standards.

Significant success has been achieved in ensuring the international acceptability of these standards; however, the applicable standard varies with the stage of the product life cycle. For example, GLP typically applies at the development stage (pre-market testing) where data are being generated for registration purposes while ISO Guide 25 would typically apply in routine post-registration regulatory (e.g. residue monitoring) or quality control activities.

An increasing number of analytical laboratories are involved in both pre- and post- registration testing activities, which result in these facilities being confronted with the need to cope with two or more of these standards. This can result in a significant burden for a laboratory as it involves separate audits and inspections, often by different organizations which typically operate on a fee for service basis.

There are two distinct, although related, issues to consider in this regard; the introduction of a unified approach to monitoring adherence to the individual quality standards and the potential for integrating the standards into a single uniform document. Although the latter scenario would be desirable, it is recognized that the greatest short-term potential for harmonization rests with the issue of integrated compliance monitoring.

These issues have been recognized internationally where there is a growing trend to more or less formally link the monitoring activities of the NAB with those of the GLP CMA. Even the possibility of combining the standards has already been discussed in some forums due to the appreciation that while the vehicle differs, their quality focus is consistent. These endeavours are supported by a growing number of parties with a vested interest in the issue, for example: EURACHEM (a network of European national laboratories which have an interest or responsibility for chemical analysis and related quality issues) "... would like to see and is willing to contribute to ... greater harmonisation of the terminology and requirements of ISO/IEC Guide 25, ISO 9000 series, GLP and other quality management systems leading to only one assessment" (*8*); "... As a result of discussions between the OECD Panel on GLP and EAL (European Cooperation for Accreditation of Laboratories), there is a realistic possibility of combining EN 45001 and OECD-GLP ..."(*9*).

A 1993 issue of the Valid Analytical Measurement (VAM) Bulletin (*10*) asked the pertinent questions: "... Is there a need to have several quality systems running in parallel? ... Would it be preferable to devise a single quality system which covers all aspects of chemical testing work?..". In a 288 laboratory telephone survey, 69% of the respondents believed that "... their quality needs ... would be best served by a single scheme ..". Despite the limitations of this survey, it can be argued that the results are likely representative of the analytical laboratory community in many countries and that the percentage of respondents in favour of a single scheme might even be higher today.

Many side-by-side comparisons of ISO Guide 25 and the OECD principles of GLP have been made (*11-13*), and the typical conclusion is that both have several common requirements although there are differences in emphasis. In addition, a limited number of requirements are specific to each standard.

In light of the above discussion and in a context of globalization where government agencies, in many sectors of the economy, are working to harmonize regulatory requirements, it may be appropriate for standardization bodies to consider a closer integration of some quality standards or at least to promote an integrated approach to compliance monitoring. This would seem to be in line with their mandate to facilitate international trade and reduce duplication while promoting the quality and validity of decision-oriented test data.

Production of Joint Residue Zone Maps for Canada and the United States

Canada, in consultation with the U.S., has developed residue maps that delineate regions or zones that are unique field residue trial regions that extend, in some cases, from Canada into Northern areas of the U.S. The purpose of these maps is to provide a scientific basis for determining the number and location of residue field trials necessary for both full and minor use registration of pesticides in both countries. The maps allow for residue data produced at any location within zones to be used as supporting data for registration thus avoiding duplication of requirements to meet national needs. This provides both a cost saving to the industry and farming community and a closer linking of Canadian and U.S. residue data requirements.

Overview of the Method for Delineating Crop Field Trial Regions. As a first step prior to the delineation of the Canadian regions, the geographic descriptions provided in the EPA document "Pesticide Reregistration Rejection Rate Analysis, Residue Chemistry"(*14*), were used to digitize the U.S. crop field trial regions. This work was completed in order to ascertain problems that might be associated with the delineation of the Canadian regions at the U.S./Canada border.

Canadian zone maps have then been identified utilizing available Geographic Information System (GIS) data and are based on agricultural ecumene and agricultural land use. The use of these data was correlated with GIS data on ecozones, soil maps, vegetation cover, ecoclimate regions, and climate data to determine delineation boundaries for unique zones or regions. The zone maps produced by this analysis have been verified through a preconsultative mechanism with both U.S. EPA and with the pesticide industry in the U.S. and Canada. The purpose of this preconsultation was to ensure the scientific validity of the analysis.

Five maps depicting the field trial regions throughout Canada, the Northern U.S. and North America have been produced in digital and hardcopy colour format. On one of these maps, identified as the "Canadian and U.S. Major and Minor Crop Field Trial Regions", both countries have been divided into a total of 18 zones and subzones. Each of these zones recognizes physical characteristics, such as soils, crops and climate, which make the region unique within the Canadian and American agricultural landscape. The subzones address differences within a region, generally reflected in the types of crops grown in that region. While some zones are specific to each country, others extend on both sides of the border.

A more detailed description of the methodology used to delineate the zones, as well as all the maps produced can be found in section 9 of the PMRA Residue Chemistry Guidelines document (3) or on the PMRA internet site at http//www.hc-sc.gc.ca/pmra.

Next Steps. A project has recently been initiated to delineate field trial regions in Mexico using the same criteria as were used for Canada. This project involves consultation with all parties involved to ensure mutual acceptance of the zone maps between the three countries.

Conclusion

The PMRA is very active in a wide range of international harmonization initiatives both under NAFTA and the OECD Pesticides Programme. Some of the specific projects the PMRA is involved in include the harmonization of data requirements in a number of areas, joint review and work sharing initiatives, development of guidance documents for industry data submission, and country data review reports, as well as for a uniform approach to data interpretation, the development of joint residue maps covering both Canada and the U.S., the development of a unified approach for electronic data submission, and the introduction of GLP.

The PMRA is committed to play an active role in harmonizing pesticide registration processes internationally while maintaining Canadian health and environmental standards. It is felt that the international context is more conducive than ever to rapid and concrete progress towards "practical" harmonization where international guidelines/procedures/requirements are routinely implemented at the national level. For example, the fact that many countries are faced with the need to reevaluate large numbers of "older" pesticides, while at the same time facing budgetary constraints, should promote a cooperative approach to registration in an effort to leverage the limited available resources.

Acknowledgments

I would like to express my sincere thanks to many colleagues in PMRA for their review and comments during the preparation of this paper.

Literature Cited

1. Chemistry Requirements for the Registration of a Technical Grade of Active Ingredient or an Integrated System Product, Pro97-01; Health Canada, Pest Management Regulatory Agency, Ottawa, 1997.
2. Chemistry Requirements for the Registration of a Manufacturing Concentrate or an End-Use Product. . . Pro97-02; Health Canada, Pest Management Regulatory Agency, Ottawa, 1997.
3. Residue Chemistry Guidelines, Pro97-03; Health Canada, Pest Management Regulatory Agency, Ottawa, 1997.
4. The OECD Principles of Good Laboratory Practices, Environment Monograph No. 45; Organisation for Economic Co-operation and Development, Paris 1992.
5. Good Laboratory Practices, Pro96-02; Health Canada, Pest Management Regulatory Agency, Ottawa, 1996.
6. Revised Guides for Compliance Monitoring Procedures for Good Laboratory Practice, Environment Monograph No. 110; Organisation for Economic Co-operation and Development, Paris 1995.
7. The Use of Laboratory Accreditation with Reference to GLP Compliance Monitoring: Position of the OECD Panel on Good Laboratory Practice (as endorsed by the 22nd Joint Meeting of the Chemicals Group and Management Committee of the Special Programme on the Control of Chemicals on 16th November, 1994).
8. De Bievre, P.; Kaarls, R.; King B. *Accreditation and Quality Assurance* 1997, *vol.* 2, pp. 89-91.
9. Pfannhauser, W. *Accreditation and Quality Assurance* 1997, *vol.* 2, pp. 101-102.
10. King, B. *VAM Bulletin* 1993, no. 10, pp. 3-4.
11. Van't Klooster, H.A.; Deckers, H.A.; Baijense, C.J.; Meuwsen, I.J.B.; Salm, M.L. *Trends in Analytical Chemistry* 1994, *vol.* 13, no. 10 pp. 419-425.
12. International Guide to Quality in Analytical Chemistry. An Aid to Accreditation; Co-Operation on International Traceability in Analytical Chemistry (CITAC), English edition 1.0, December 1995, ISBN 0-948926-09-0.
13. Platt, S.F. *VAM Bulletin* 1993, no. 9, pp. 7-10.
14. Pesticide Reregistration Rejection Rate Analysis, Residue Chemistry, USEPA, OPTS (H-7508W), EPA 738-R-92-001, June 1992.

Chapter 24

Quality Assurance for Environmental Laboratories in Canada

Richard Turle[1], Neil McQuaker[2], and Rick Wilson[2]

[1]Environmental Technology Centre, Environment Canada, Ottawa, Ontario K1A 0H3, Canada
[2]Canadian Association of Analytical Laboratories, Suite 300, 265 Carling Avenue, Ottawa, Ontario K1S 2E1, Canada

Canada has developed a national quality assurance system, based on ISO Guide 25, to provide accreditation and certification to laboratories that provide analytical systems data. The certification for key environmental parameters is based on the analysis of proficiency samples. This system involves two partners, the Standards Council of Canada as the accrediting body and the Canadian Association of Environmental Analytical Laboratories as the program provider. This system has been effective in meeting the demands of regulators and commercial clients.

There are a number of characteristics of a quality management system for laboratories engaged in primarily routine analysis that are essential to its optimal design and successful implementation. Such characteristics may in some aspects be different from those laboratories engaged in GLP protocol driven testing where the effort is directed to exhaustive testing of a single compound or product. Key among these essential characteristics is that the system should apply to all types and sizes of laboratories. Many laboratories attached to field testing stations or industrial plants are often quite small and cannot afford a full time quality assurance officer, for example. Other important characteristics are as follows:

- The system should apply equally to both private and public sector laboratories. This ensures that there will be no hiding within the bureaucracy of a laboratory that cannot meet the requirements of the quality system. For example, the system may allow that in public sector laboratories the need for confidentiality is less important than in a private sector laboratory, especially one under an Access to Information Act or similar legislation.

- The system should be recognized nationally, in all jurisdictions and by all levels of government, industrial trade associations and professional bodies. Such a national system should also follow internationally recognized standards.

- Finally, any laboratory engaged in routine testing should also participate in proficiency testing schemes based on, as much as possible, using real samples.

Accreditation. Many governments and other bodies run accreditation schemes for a variety of purposes including laboratory performance. In today's world, where environmental and regulatory decisions demand data of high and known quality, accreditation gives laboratories the recognition that they are capable of producing quality data on the tests described in their scope of accreditation.

What are the requirements for accreditation? Summarized, they are:

- a full or part time quality assurance officer;
- a quality manual, describing the laboratory and outlining policies
- a methods manual containing all routine procedures
- Standard Operating Procedures, including those for modification of methods, corrective procedures, and non-conformances to the quality manual
- participation in proficiency testing (certification) schemes and other inter-laboratory comparisons (round-robins) to demonstrate competency to perform routine tests
- a site inspection or audit every two years.

Certification. Certification is recognition that a laboratory has actually analyzed blind samples and has met the requirements in terms of both accuracy (bias) and precision. Generally, a certificate is awarded for either a single test (e.g., pH, total PCBs) or for a group of tests (e.g., PAHs in soil, anions on air filters). A good certification program will demand that blind samples be run at least twice a year. The submitted results are compared to either a reference value, if a Certified Reference Material is used, or to a consensus mean. Points can be assigned on the basis of "acceptable deviation" from the accepted value. A score of 70% or more is required on two successive rounds to maintain the certificate of proficiency. Coupled with accreditation, certification provides further confidence that a laboratory can indeed perform quality test procedures. It is possible in Canada to be certified without being accredited but experience indicates that most laboratories which enter the certification program receive accreditation within two years.

Perspectives on Laboratory Accreditation

There are many perspectives on the value of lab accreditation depending on whether one is a laboratory practitioner or a user of laboratory data.

The Laboratory Manager's Perspective. The introduction of a quality system brings about a "cultural change" into the laboratory. Apart from the obvious pride in obtaining the certificate of accreditation, there is a change in attitude. This leads to doing it right the first time. This attitude also accepts that errors - nonconformances in the language of accreditation - will occur but that the lesson will be learned or the situation will be corrected. Inevitably, this reduces errors to the absolute minimum. This pride in doing the job right leads to a continuous improvement in procedures. Since procedures must be evaluated at least every two years. Consistency is thus ensured because no change is made without consultation and review. Further, written procedures can help define, hasten and improve laboratory training programs. All of these improvements lead to both tangible and intangible benefits for the productivity-conscious laboratory manager.

The Commercial Laboratory's Perspective. The possession of accreditation leads to more business opportunities whether or not the testing is required by a regulator for environmental assessments, contaminated site clean up, or for other commercial purposes. Accreditation and, if appropriate, the participation in a suitable proficiency testing program increasingly is written into contracts to testing organizations by private companies as well as governments.

The Laboratory Client's Perspective. Accreditation provides to the client a third party assessment of the laboratory's capability and performance. The proficiency testing reports indicate whether the laboratory can perform the desired tests competently. The site assessment reports will identify problem areas and also indicate steps the laboratory has taken over and above that required to obtain accreditation. Finally businesses, who have to satisfy ISO 9000 requirements, can do this by using an accredited laboratory meeting ISO Guide 25 standards.

The Regulator's Perspective. The regulator, by insisting that only accredited and, where appropriate, certified laboratories perform testing for a given regulation, is defining that as a minimum standard for the group of laboratories so affected. This gives him the assurance that laboratories operating out of kitchens or garages will not be providing vital results. He will know that the laboratories are using written test procedures and that there is an applicable set of Standard Operating Procedures. This reduces the cost of compliance auditing and inspection of laboratories.

International Recognition

The key to international recognition for laboratory accreditation is ISO Guide 25 (*1*). ISO Guide 25 is increasingly used as a basis for analyses required for trade. Both the Canadian and Mexican accreditation systems are based on it,as well as the developing National Environmental Laboratory Accreditation Conference (NELAC) program of the USEPA (*2*). Analytical data coming from a laboratory accredited to ISO Guide 25 standards will meet minimum defined standards for quality. This results in acceptance of data between governments and across borders. As environmental problems are becoming more global in scope, this will become even more essential.

The Canadian Association of Environmental Analytical Laboratories (CAEAL)

CAEAL is a not-for-profit organization which exists for the purpose of promoting quality in environmental laboratories. The need for such an organization was recognized by both public and private laboratories to dispel the notion that laboratories were generating data of disparate quality. The only solution was to require laboratories to produce verifiable data of laboratory performance within an accreditation scheme. Established in 1989, the organization has grown and developed to become recognized internationally as an example of an effective national accreditation body. The Board consists of representatives elected from both government laboratories (including federal, provincial and municipal) as well as the private (for-profit) sector. In most jurisdictions in Canada there is no legal requirement to be accredited though at least one province, Newfoundland, requires it for results submitted to the provincial Department of the Environment. Other jurisdictions are actively considering accreditation as a requirement.

CAEAL elements. This is a national program, open to laboratories in every province. However, in the province of Quebec, the Ministère del l'environnment et faune operates its own scheme (*3*), which also meets ISO Guide 25 requirements. Although the CAEAL program is based on ISO Guide 25, laboratories also must meet the requirements of a unique Canadian standard for environmental analytical laboratories, namely CAN/CSA-Z753-95 (*4*). The program requires that laboratories must also participate in CAEAL and other proficiency testing programs if they are applicable to the laboratories' scope of testing. There is also a requirement that each laboratory receives a site audit which comprises an examination of the quality manual and the testing and quality assurance procedures and includes a trace of selected samples through the testing procedures. All CAEAL auditors are trained to international standards. About 90 laboratories are now part of the SCC/CAEAL accreditation program. This constitutes about half of all the laboratories accredited in Canada. These include most federal and provincial laboratories and a few municipal laboratories. Most private sector environmental laboratories outside of Quebec have received accreditation through SCC/CAEAL. The growth is now occurring in the industrial sector as businesses recognize that their own laboratories should meet international standards.

CAEAL certification program. The program consists of chemical analyses on a variety of sample types but mainly water. The program covers analyses for major cations, anions and trace metals in water and on air filters, organochlorine pesticides, polycyclic aromatic hydrocarbons, volatile organic compounds in water, PAHs in soil, and PCBs in oil, as well as coliform, daphnia, trout and Microtox toxicity tests. Four samples are analysed twice a year. A scheme of acceptable deviation from the reference or consensus mean is calculated. A perfect score gives a rating of 100. To maintain the laboratory's status in the program a score of no less than 70 must be attained in each round of testing. Currently there are over 160 laboratories participating in this program.

Environment Canada. This is the environmental agency of the Canadian federal government. It has supported CAEAL since inception by providing auditors and technical assistance. It was felt by senior management that by supporting a non-governmental laboratory organization, it could be eventually self financing, which has indeed happened. Further, the amount of work required for compliance activities would be reduced if Environment Canada could be assured that the laboratory community as a whole was producing reliable data. This again has indeed happened. Similar support was obtained from the Ontario Ministry of Environment and Energy for similar reasons. Environment Canada requires accreditation for laboratory work it contracts to the private sector. Further, its operational laboratories are accredited.

Standards Council of Canada and CAEAL

The SCC is the body legislated by the Canadian parliament to develop national standards in Canada. It represents Canada at the International Standards Organization (ISO). It accredits laboratories under the Program for Accreditation of Laboratories in Canada (PALCAN) which includes those recommended by CAEAL. The SCC is starting to develop mutual recognition agreements with other national organizations. In this regard, there is a strong desire by Canadian companies and governments to see a North American-wide Mutual Agreement to complement the work undertaken by NAFTA. The SCC and CAEAL signed an agreement to enhance and develop the initial program. CAEAL recommends laboratories to the SCC which meet the CAN/CSA Z753. The SCC then grants accreditation to environmental laboratories for the tests they have specified in their application. The SCC promotes the use of accredited laboratories by publishing a list of laboratories biannually (*5*). This unique partnership has strengthened both organizations. It has meant that laboratories which had to be accredited by both organizations can now receive a joint site inspection and a single certificate of accreditation. The SCC benefits by having a larger pool to draw on for input into new or revised standards such as ISO Guide 25. For CAEAL, it will mean that there will be no need to develop its own mutual recognition agreements once the SCC develops such agreements.

What of the future?

It is anticipated that there will be a slow but gradual increase in the number of laboratories accredited under the SCC/CAEAL program. Growth is most likely in the municipal and industrial sectors. We anticipate expansion of the certification program as it becomes possible to add suitable chemical parameters (e.g., mercury arsenic and selenium). There is a need for a dioxin proficiency sample program. We also see collaboration between CAEAL and regulating bodies such as Environment Canada to develop methods such as one for all types of petroleum hydrocarbons in contaminated soil.

Literature Cited

1. ISO Guide 25: 1990, *General Requirements for the Competence of Calibration and Testing Laboratories.*
2. *Analytical Chemistry News and Features* , October 1, 1997, 589A.
3. *Programme d'accreditation des laboratoires d'analyse environnementalle*, Quebéc Ministère de l' environnement et de la faune, 1994.
4. *CAN/CSA-753-95, Requirements for the Competence of Environmental Laboratories*, Canadian Standards Association, Rexdale, (Toronto) 1995.
5. *Directory of Accredited Calibration and Testing Laboratories, CAN-P-1550*, Standards Council of Canada, 1996.

Chapter 25

Registration Procedures for Agrochemicals in the European Union

Jorge-I. Celorio

Hoechst Schering AgrEvo GmbH, Building K607, D–65926 Frankfurt—AM—Main, Germany

With Directive 91/414 the European Union has achieved a harmonized approach for the registration of plant protection products. The procedures foresee evaluation and registration in two phases: 1) assessment of an active substance and recommendation by a Rapporteur Member State; 2) registration in the Member States of the formulated products containing the active substance, if it has been accepted by the Standing Committee on Plant Health. If rejected, the active substance will be banned in the EU. For reregistrations, the active substances to be reviewed are published by the Commission. All companies interested in defending the registrations of a given active substance must notify their intention. For both new and old compounds, the notifiers have to provide the Rapporteur Member State with the corresponding dossier. The assessment by the Rapporteur Member State leads to the production of a draft Monograph, which is discussed within expert groups and includes the recommendation of acceptance/rejection. This leads to the voting within the Standing Committee on Plant Health.

For years the European Union had been striving to achieve the harmonisation of registration procedures for plant protection products in all Member States. These efforts crystallized in 1991 with the publication of Directive 91/414 (*1*) — which laid the base for a common procedure for the (re)registration of old and new products — with the goal of reviewing all existing active substances in the EU (over 700) in twelve years. This Directive was followed by a series of Regulations, Directives, Guidelines, etc., which demonstrate how complex the European agricultural situation is, affected by climatic, political and cultural factors which differ radically between the northern and southern Member States. The intention of this paper is to give an overview of the current EU procedure.

Directive 91/414/EEC

Directive 91/414 (*1*) is titled "concerning the placing of plant protection products on the market" and sets the basic procedures for accomplishing this. Basically, it foresees the registration of plant protection products in two phases. In the first phase, the assessment of an active substance (a.s.) is allocated by the Commission to a Member State. The designated Rapporteur Member State (RMS), after reviewing the documentation presented, makes a recommendation to the Commission (favorable or unfavorable). At the Standing Committee on Plant Health (SCPH) a vote is held on the recommendation, and if a certain majority is obtained, the a.s. will be listed in the so-called 'positive list', the Annex I to Directive 91/414. Once the compound is in Annex I, the submitter can proceed with the second phase, which is the registration at the national level of the plant protection product. If the a.s. cannot be included in Annex I, it will be banned in the entire EU, and all Member States must revoke the registrations of plant protection products containing this active substance.

Directive 91/414 contains six annexes, as shown below:

Annex I	Positive list of evaluated and accepted active substances
Annex II	Data requirements for the active substance
Annex III	Data requirements for formulation(s)
Annex IV	Risk phrases
Annex V	Safety phrases
Annex VI	Uniform principles

The data requirements are described below (cf. Documentation). The Uniform Principles contain guidance for the evaluation and decision making concerning the submitted data, including trigger values for acceptance/rejection and for demanding further studies.

The Commission Directive 91/414 has been followed by at least ten Directives, five Regulations, eighteen Decisions and twenty-three Guidelines/Working Documents (Tables I and II). Commission Regulation 3600/92 (*2*) laid down "the detailed rules for the implementation of the first stage of the programme of work referred to in Article 8(2)" of Directive 91/414. This Regulation, which listed the first 90 substances to be reviewed, was subsequently amended by Regulations 933/94 (which listed the substances to be assessed and their distribution among 12 Member States) (*3*) and 491/95 (which integrated the new Member States of Austria, Finland, and Sweden in the review process and redistributed the remaining 87 active substances among all 15 Member States) (*4*) (Table III).

Although the intended goal was to review all existing active substances in the EU (over 700) in twelve years, this goal cannot be achieved, since the amount of work and the complexity of certain issues (e.g., data protection) were grossly underestimated. Even the basic philosophy toward agrochemicals in general can lead to controversial views within the EU for both new and old substances. In some Member States in Northern Europe, agriculture plays a secondary role, representing maybe less than 5% of the gross national product (GNP). In Southern Europe, however, agriculture may represent 40% of the GNP. This taints politically the

Table I. EU Documents: Directives, Decisions and Regulations

Document No.	Context
Dir. 91 / 414	Basic registration directive
Dir. 93 / 71	Efficacy Data
Dir. 94 / 37	Identity and Physicochemical Data
Dir. 94 / 43	Establishes the Uniform Principles (annulled)
Dir. 94 / 79	Toxicology, Metabolism and Operator Exposure
Dir. 95 / 35	Derogates GLP from testing on beneficials and for residues
Dir. 95 / 36	Environmental Fate
Dir. 96 / 12	Ecotoxicology
Dir. 96 / 46	Analytical Methods
Dir. 96 / 68	Residues
Dir. 97 / 57	Establishes the (new) Uniform Principles
Reg. 3600 / 92	Establishes the rules and lists the first 90 active substances
Reg. 933 / 94	Allocation of 89 a.s. among 12 Rapporteur Member States
Reg. 491 / 95	Re-allocation of 87 a.s. among 15 Rapporteur Member States
Reg. 2230 / 95	Re-establishes deadline for the submission of dossiers
Reg. 1199 / 97	Amends Reg. 3600/92
Dec. 94 / 643	Cancellation of cyhalothrin registrations in the EU
Dec. 95 / 276	Cancellation of azinphos-ethyl and ferbam registrations in the EU
Dec. 96 / 266	Recognizes completeness of the dossier of kresoxim-methyl
Dec. 96 / 341	Recognizes completeness of the dossier of flurtamone
Dec. 96 / 457	Recognizes completeness of the dossier of quinoxyfen
Dec. 96 / 520	Recognizes completeness of the dossier of prohexadione-calcium
Dec. 96 / 521	Recognizes completeness of the dossier of chlorfenapyr
Dec. 96 / 522	Recognizes completeness of the dossier of spiroxamine
Dec. 96 / 523	Recognizes completeness of the dossier of azoxystrobin
Dec. 96 / 524	Recognizes completeness of the dossier of isoxaflutole
Dec. 96 / 586	Cancellation of propham registrations in the EU
Dec. 97 / 137	Recognizes completeness of the dossiers of prosulfuron and cyclanilide
Dec. 97 / 164	Recognizes completeness of the dossiers of azimsulfuron, flupyrsulfuron-methyl and *Paecilomyces fumosoroseus*
Dec. 97 / 248	Recognizes completeness of the dossier of *Ps. chloroaphis*
Dec. 97 / 362	Recognizes completeness of the dossiers of carfentrazone-ethyl, fluthiamide and fosthiazate
Dec. 97 / 579	Establishing Scientific Committees in the fields of consumer health and food safety (DG XXIV)
Dec. 97 / 591	Recognizes completeness of the dossiers of ethoxysulfuron, mefenoxam, famoxadone and *Ampelomyces quisqualis*
Dec. 97 / 631	Recognizes completeness of the dossier of flumioxazine

Table II. EU Documents: Guidelines and Working Documents

Document No.	Context
2949 / VI / 93	Article 12
9016 / VI / 93	Overview of the state of main works in DG VI
1654 / VI / 94	Preparation of Monograph by Rapporteur Member State
1663 / VI / 94	Preparation of dossiers
7109 / VI / 94	Applicability of GLP
1606 / VI / 95	Contact addresses
1614 / VI / 95	Working procedure for 'old' active substances
1642 / VI / 95	Ecotoxicology
1663 / VI / 95	Working procedure for new active substances
1694 / VI / 95	Modeling Environmental Fate: Leaching
4952 / VI / 95	Modeling Environmental Fate: Leaching
4992 / VI / 95	Microorganisms and viruses
4993 / VI / 95	Microorganisms and viruses
7017 / VI / 95	Acceptability of data re GLP
7027 / VI / 95	Completeness check of dossiers
7531 / VI / 95	Setting acceptable operator exposure levels
7600 / VI / 95	Biological assessment dossier (Efficacy data)
4754 / VI / 96	Data protection
6476 / VI / 96	Modeling Environmental Fate: Surface water
7617 / VI / 96	Modeling Environmental Fate: Soil
8538 / VI / 96	Draft on completeness check of dossiers
1607 / VI / 97	Generation of residue data
1635 / VI / 97	Inclusion of diquat in Annex I

decision-making process and leads to compromises and delays. Besides, the EU is formed by 15 independent states, and EU guidance may be influenced at the national level by local legal restrictions.

The Registration Procedure: New Active Substances

The registrant notifies the Commission through a Member State of the intention to register a new a.s. The Commission then allocates the review of the a.s. to a Rapporteur Member State, which may be the one through which the notification was made. The notifier submits the data dossier (*5*) and the RMS proceeds to make the completeness check and a preliminary evaluation.

Once the Commission has confirmed that the submitted dossier is complete, the Member States may issue Provisional Registrations for use of formulated products containing the new active substance. These Provisional Registrations permit the marketing of a product in a given Member State for a maximum of three years. Until October 1997, Provisional Registrations had been granted to the following active substances: azimsulfuron, azoxystrobin, flupyrsulfuron-methyl, flurtamone, kresoxime-methyl, quinoxyfen, and spiroxamine.

A complete evaluation of the submitted data is carried out by the RMS (*6*). If necessary, additional data may be required. The RMS finally prepares a Monograph summarising the principal characteristics of the active substance and its risks and benefits and makes a recommendation of inclusion in or exclusion of Annex I of Directive 91/414

The Registration Procedure: Reregistration of Old Active Substances

A list of ca. 90 active substances to be reviewed was prepared by the Commission and published as an EU document (*2*). All organisations which marketed products containing these active substances had to notify the Commission whether they were interested in reregistering their a.s. at the EU level. A list of notifiers was then published (*3*), so that the organisations could meet and decide whether they would eventually form a 'Task Force' for a given a.s. and submit jointly one dossier.

The Commission also allocated the active substances among the Member States (*3*). The notifiers submitted to the corresponding RMS the data relevant to their product. This consisted of the dossiers described below (cf. Documentation).

As for new substances, the RMS evaluates the data, ending with a Monograph and a recommendation of inclusion in or exclusion from Annex I of Directive 91/414.

Decision-Making

The Monograph consists of four volumes (*6*):

- Volume 1: statement on subject matter, conclusions, decisions and proposals, and further information;
- Volume 2: list of studies submitted;
- Volume 3: summary, evaluation and assessment;
- Volume 4: confidential information.

The Monograph, the Tier 2 and 3 summaries (cf. Documentation), and, if need be, the whole dossier are reviewed by all Member States. The Monographs are evaluated and discussed by sections within the European Community Co-ordination (ECCO) expert groups, which consist of 5 experts on each field from the Member States. The first ECCO meetings were organised by the British Pesticides Safety Directorate (PSD) in York and the German Biologische Bundesanstalt (BBA) in Braunschweig. In these meetings the RMS — author of the Monograph — defends the results and the conclusions described in the Monograph. Once the review is finished, a report is sent to the Commission. Based on the RMS's Monograph and recommendation, the ECCO report, and inputs from the Directorates General VI (Agriculture), XI (Environment) and XXIV (Food and Consumer Safety) — inputs which could veto the inclusion in Annex I — a vote is taken at the SCPH, where again the RMS defends, at a political level, the proposal found in the Monograph. Of a total number of votes of 87, a minimum of 62 votes in favor of inclusion in Annex I is needed if the a.s. is to be used in the EU. If 26 votes are against the inclusion, the a.s. has to be prohibited in the EU following Directive 79/117 (7). The distribution of votes among Member States is shown in Table III.

Table III. Distribution of Active Substances among the Member States of the EU and of Votes in the Standing Committee on Plant Health.

Country	No. of A.S.[1]	No. of Votes in SCPH[2]
Austria	4	4
Belgium	5	5
Denmark	3	3
Finland	3	3
France	11	10
Germany	11	10
Great Britain	11	10
Greece	5	5
Ireland	3	3
Italy	11	10
Luxembourg	1	2
Netherlands	5	5
Portugal	5	5
Spain	7	8
Sweden	4	4

(1) Comm. Reg. 3600/92 (as amended by Comm. Regs. 933/94 and 491/95)

(2) Qualifying majority: 62. Blocking minority: 26. Total: 87.

If the vote within SCPH is favorable to the inclusion of the a.s. in Annex I, a Directive amending this annex of Directive 91/414 will be published. Then, the notifiers can proceed with the second phase, which is the (re)registration of the formulated product at Member State level. They have 18 months to implement any restrictions on usage and — in the case of old active substances — 5 years to re-register the corresponding products.

If the vote is not favorable to the inclusion in Annex I, the a.s. will be banned in the EU. For old substances, all registrations in the EU have to be revoked and existing stocks of products containing this a.s. have to be sold and used within 2–3 years.

Documentation

The review process asks for an enormous amount of documentation in the form of dossiers for both the a.s. (the so-called Annex II dossier) and a representative formulation (the so-called Annex III dossier). These dossiers are organized in 'Documents' lettered from A to N (Table IV) (*5*): Documents A to J and N are common to both the active substance and the formulation, while Documents K to M are prepared for each dossier with a series of summaries (the so-called 'Tier' summaries) building a 'pyramid' of information.

Table IV. EU Dossier: Individual Documents

Document	Context
A	Statement of context
B	Collective dossiers (Task Forces)
C	Existing national labels
D	Summary of authorized uses in the EU
E	Good Agricultural Practice data
F	Notification following Commission Regulation 3600/92
G	Statement on permited inerts
H	Safety Data Sheets for inerts
I	Other toxicological/environmental data
J	Confidentiality statement
K	Individual study reports (Annexes II and III)
L	Tier 1: Quality check of individual studies
M	Tier 2: Summary and evaluation
N	Tier 3: Overall assessment

The most complicated and time-consuming part is the preparation of Document K (formed by the individual study reports) and Document L (the individual quality checks, also called Tier 1 summaries). Of much more interest are Documents

M and N (Tiers 2 and 3, respectively), which summarize the scientific and technical aspects and assess the risks and benefits of the active substance and the formulation.

Documents K to N are divided into Sections (Table V), which differ somewhat for the a.s. (Annex II) and the formulated product (Annex III).

Table V. EU Dossier: Data Requirements

Section	Annex II (Act. Ing.)	Annex III (Form.)
1	Identity	Identity
2	Physicochemical Data	Physicochemical Data
3	Further Information	Data on Application
4	Analytical Methods	Further Information
5	Toxicology and Metabolism	Analytical Methods
6	Residues	Efficacy Data
7	Environmental Behavior	Toxicology
8	Ecotoxicology	Residues
9	Summary/Evaluation Sec. 7 and 8	Environmental Behavior
10	Classification and Labelling	Ecotoxicology
11	Dossier, Annex III	Summary/Evaluation Sec. 7 and 8
12	———	Further Information

The Annex II dossier includes sections on Identity, Physicochemical, Analytical, Toxicology and Metabolism, Residues, Environmental Fate, and Ecotoxicology data, plus Classification and Labelling. The Annex III dossier — which may refer to the Annex II data, especially where residues are concerned — includes also a section on Biological Efficacy.

Most of the studies carried out are conducted following internationally recognised guidelines like OECD, EPPO, FAO/WHO, EPA, etc.

Advantages / Disadvantages

After many years of trying the European Union has finally established a standardized procedure for the registration of plant protection products. The guidelines are uniform for all the Member States, and once the Uniform Principles have been established (*8*), there will also be standard criteria for the evaluation of the products.

Also, the Tier 2 and 3 summaries are structured in such a way that they can be used world wide for registration purposes. The OECD is planning in 1998 to publish guidelines based on the EU dossier format. This means that in the near future — if enough trust is built among the OECD Member States — one dossier would suffice for world-wide registration. If, e.g., the submission takes place in Canada and a Monograph is prepared by that country's regulatory authorities, this Monograph could be used by all other OECD countries, including the EU, for regulatory purposes, thus reducing the time for decision-making, workload and costs.

The disadvantages lie not in the idea but in its current implementation. The EU system is too complex, bureaucratic, and expensive. The decision-making process involving so many different entities is unclear, slow and burdensome and is not based uniquely on scientific reasoning but also on economic and especially political factors — countries like Denmark and Sweden with reduced agriculture are politically motivated to ban agrochemicals — and thus there may be no objectivity when voting at the SCPH.

Both the farmer and the agrochemical industry are also at a disadvantage. The farmer will have to wait longer for the availability of new products and may have fewer options in the future. For the industry, the system means increased costs, longer times to obtain registration and thus loss of income. The system also stretches the resources of both industry and officials to the limit or beyond.

Costs and Time

Although it is difficult to give exact costs (which depend on many changing factors), some idea about the actual costs for the obtainment and defense of registrations can be given:

Preparation of dossiers:	\$ 400000.00
Further studies:	\$ 700000.00
Fees:	\$ 100000.00
Total per active substance:	\$ 1200000.00

Between notification and inclusion in Annex I, the time interval has to be calculated in years. After notification, the notifier has 12 months to submit the data to the RMS. The evaluation of the data and the preparation of the Monograph should theoretically take place within 12 months after receiving the dossier. Then a further 3 months are necessary before the SCPH can vote on the compound. In theory, therefore, inclusion in Annex I could be achieved in less than two and a half years. In practice, these time frames were grossly underestimated.

As of November 1997 — i.e. 6 years after the directive was published and 4 years after being enforced — only one substance of the original 90 found in the first list has been included in Annex I: imazalil, a fungicide. Two further substances, the herbicide diquat and the insecticide fenthion, will probably be included in the near future. No new active substance has been included either. On the other hand, 6 substances have been or will be withdrawn from the market: azinphos-ethyl, cyhalothrin, dinoterb, fenvalerate, ferbam and propham.

Before Directive 91/414, registration times among the Member States varied between 11 and 60 months. With Directive 91/414, the registration time has not yet been harmonized, but it seems it will not be under 48 months, if the time needed for the second phase is included. For new registrations, this is unacceptable. The whole process — from submission of the dossier to the RMS to the registration of the plant protection product at the national level — should not take more than 30 months.

Conclusions

Directive 91/414 was a big step forward in harmonizing the registration requirements and procedures within the European Union. However, although the structure of the dossiers is basically sound, the practical implementation of the Directive has stressed both industry and authorities to the limit of their resources. This has led the Commission to realize that the original goal of reviewing all active substances in 12 years cannot be achieved. Also, the complexity of the system — mainly decision making, but also unresolved issues like data protection, mutual recognition, minor crops, and parallel imports — make uncertain the future of the system, especially for old products. A decision on this issue may fall within the year 1998.

After all the effort and cost, three possibilities exist. First, keep the system, but overhaul it by streamlining and agilizing it so that the registration of plant protection products — especially new products — can take place more rapidly with less bureaucracy, with quicker decision making, trust among Member States, mutual recognition, and harmonization. The second possibility would be to create a central European authority for the scientific evaluation and/or registration of plant protection products — similar to the European Medicines Evaluation Agency (EMEA) for pharmaceuticals, located in London — for phase I registration (inclusion of the active substance in Annex I) and keep national jurisdiction for phase II (registration of the formulated product). The third possibility would be to return to national legislation, which in all probability would lead to still more chaos.

Literature Cited

1. *Council Directive 91/414/EEC*, Official Journal of the European Communities No. L 230, 19.8.91, p. 1, as corrected in Official Journal of the European Communities No. L 170, 25.6.92, p. 40.
2. *Commission Regulation 3600/92*, Official Journal of the European Communities No. L 366, 15.12.92, p. 10.
3. *Commission Regulation 933/94*, Official Journal of the European Communities No. L 107, 28.4.94, p. 8.
4. *Commission Regulation 491/95*, Official Journal of the European Communities No. L 49, 4.3.95, p. 50.
5. *Working Document 1663/VI/94 rev. 6*, 24.5.95.
6. *Working Document 1654/VI/94 rev. 6.2*, 2.2.97.
7. *Council Directive 79/117/EEC*, Official Journal of the European Communities No. L 33, 8.2.79, p. 36.
8. *Proposal for Commission Directive 97/C 240/01*, Official Journal of the European Communities No. C 240, 6.8.97, p. 1.

Chapter 26

GLP in the European Union: Regulations, Implementation, and Experiences

Jutta Lange[1] and Jorge-I. Celorio[2]

[1]Schering Aktiengesellschaft, Berlin, Germany
[2]Hoechst Schering AgrEvo GmbH, Building K607, Frankfurt—AM—Main, Germany

With Directive 87/18/EEC the European Union officially requested its Member States to implement at the national level the OECD GLP Principles of 1981. This has been carried out between 1986 and 1993. Unfortunately, differences in the dates and organisation of implementation, interpretation of the Principles, and monitoring practices have led to discrepancies between the Member States. In order to reduce these differences, the "Mutual Joint Visits" among the Member States were introduced. It is still to be seen whether this will lead to harmonization. An overview of monitoring authorities, of the number of inspections carried out, of the number of facilities certified, and of some of the specific findings in several Member States of the European Union are given.

In December 1986 the European Union (EU) published Directive 87/18 (*1*), by which its Member States should implement the OECD GLP Principles of 1981 (*2*) within two years. This directive was followed by Directives 88/320 of June 1988 (*3*) and 90/18 of December 1989 (*4*) which require the establishment of a national monitoring program in each state of the Union. All three Directives laid the base for the implementation of the OECD GLP Principles in the EU. Unfortunately, these Directives have been implemented in all of the EU at different times and with different systems. Thus, there are countries with several years' experience beside countries with little or no experience with these issues. As a consequence, the EU conducted the "Mutual Joint Visits" (MJV) program in 1995 and 1996, in order to evaluate these differences with the hope that these visits will eventually lead to measures that will harmonize status and procedures.

EU Directives

Directive 87/18 came into force in 1989. This Directive demands that Member States of the European Union establish legal requirements asking for compliance with the OECD Principles of GLP in the conduct of safety studies for the (re)registration and marketing of agrochemicals, pharmaceuticals, and chemicals in general.

The Member States of the EU reacted differently. Some States required GLP at a national level prior to the publication of 87/18 (e.g., Great Britain, Sweden). Other States were in no special hurry to implement this Directive. Thus, as shown in Table I, the implementation of Directive 87/18 in the EU Member States took place between 1986 (Great Britain, Italy) and 1993 (Spain). Sweden and Great Britain officially transposed 87/18 into national law in 1996 and 1997, respectively.

Table I. National Implementation of OECD GLP Principles or EU Dir. 87/18

Country	ISO Code	Year
Austria	AUT	1989
Belgium	BEL	1988
Denmark	DNK	1989
Finland	FIN	1990
France	FRA	1990
Germany	DEU	1990
Great Britain	GBR	1986[1]
Greece	GRC	1988
Ireland	IRL	1991
Italy	ITA	1986
Luxembourg	LUX	—
Netherlands	NLD	1989
Portugal	PRT	1990
Spain	ESP	1993
Sweden	SWE	1991[2]

[1] Voluntary implementation in 1986; transposition of 87/18 in 1997
[2] GLP program since 1979; transposition of 87/18 in 1996

The implementation of Directives 88/320 and 90/18 was more complex, since these Directives require the establishment of monitoring authority(ies) at a national level. This was left to the individual Member States, together with the training of the corresponding inspectors.

Monitoring Authorities

The organization of national monitoring authorities ranges from one central team to countries which — bound by national laws — had to establish local groups taken from other regulatory or monitoring activities. Thus, Great Britain has a team of *ca.* 5 experts responsible for conducting inspections in all relevant fields where GLP is applied, while Germany has 17 independent groups (16 State groups for local inspections of privately-owned test facilities, plus the Federal GLP Office for Federal and foreign facilities) with *ca.* 115 inspectors which dedicate an average of approximately 15% of their time to GLP monitoring activities. Both teams carry out approximately the same number of inspections yearly (Table II) (*5*). All other Member States lie somewhere in between.

The EU GLP Working Group in Brussels considers that a minimum of two inspections per year is necessary to maintain a certain experience in monitoring GLP (*6*). From the data shown in Table II the average number of inspections can be calculated. Nevertheless, not all EU Member States comply with this suggestion.

Mutual Joint Visits

In order to assess possible differences in monitoring practices between the Member States (cf. National Inspections), the EU started in 1995 the "Mutual Joint Visits" (MJV) program. This consisted of having three Member States as observers during the monitoring activities of a host country. Six countries were visited in 1995, nine in 1996. Luxembourg has no facilities, thus it was excluded from the program. Norway, although not a member of the EU, asked to be included.

Table III shows the host country, the visiting Member States, and the date of the visit. The underlined country was responsible for preparing the report on the visit.

The EU handles the information obtained from these visits as confidential and, therefore, no details can be given. However, in a rough way, the following differences are known (*7*):

- Interpretation and implementation of the OECD Guidance Documents Nos. 2 and 3 (monitoring)
- Documentation of the monitoring activities
- Training and experience of the official inspectors
- Archiving of documents of the monitoring activities
- Collaboration with other regulatory authorities

A final report on the MJVs was prepared based on the visit reports. This report was due to be discussed within the EU in the fall of 1997 in order to decide what should be the following steps.

Table II. GLP Inspections in the EU*

Country	Dates	Facilities Inspected	Facilities/Year	In Compliance	Pending	Not in Compliance
AUT	03/91 – 12/96	5	1	4	0	0
BEL	09/88 – 01/97	33	4	17	1	0
DEU	06/89 – 12/95	200	30	198	2	0
DNK	11/89 – 11/96	16	2	15	1	0
ESP	02/96 – 12/96	5	5	5	0	0
FIN	10/91 – 05/96	7	1	7	0	0
FRAU	12/92 – 04/96	98	29	82	7	3
GBR	01/91 – 12/96	208	35	180	13	2
GRC	10/95 – 03/97	7	5	5	2	0
IRL	05/92	1	—	1	0	0
ITA	07/88 – 01/97	43	5	43	0	0
NLD	03/87 – 11/96	52	5	36	4	3
PRT	03/95 – 10/96	2	1	2	0	0
SWE	04/92 – 11/96	18	4	18	0	0
Totals		695		613	30	8
%		100		88	4	1

* EU Annual GLP Inspection Report 1996

Table III. Mutual Joint Visits

Host Country	Visiting Countries[1]	Date
Austria	Belgium, Germany, Finland	January 1996
Belgium	Ireland, Netherlands, Norway[2]	May 1996
Denmark	Finland, Greece, Netherlands	March 1996
Finland	Portugal, Spain, Sweden	May 1996
France	Italy, Norway[2], Sweden	August 1996
Germany	Denmark, France, Spain	February 1996
Great Britain	Ireland, Netherlands, Portugal	December 1995
Greece	Austria, Great Britain, Norway[2]	October 1996
Ireland	Austria, Finland, Great Britain	April 1996
Italy	Great Britain, Portugal, Sweden	October 1995
Luxembourg[3]	———	———
Netherlands	Belgium, France, Greece	July 1995
Norway[2]	Germany, Italy, Spain	October 1996
Portugal	Denmark, France, Germany	June 1995
Spain	Austria, Belgium, Denmark	September 1996
Sweden	Greece, Ireland, Italy	August 1995

[1] The underlined country was responsible for preparing the report on the visit.
[2] Non-EU country; participated voluntarily.
[3] Has no facilities; excluded from the MJV program.

Currently, discussions are taking place within the OECD to eventually implement a similar program for its members. The Federal Republic of Germany may also introduce mutual joint visits among its 17 monitoring groups (representing the different Federal States).

National Inspections

Table II summarizes the inspections carried out by EU national monitoring authorities, as obtained from the European Commission (*5*).

The countries with the most test facilities — and, therefore, with the most inspections — are France, Germany and Great Britain. These countries examined in the corresponding time frame *ca.* 30 facilities per year, although the implemented monitoring systems are completely different.

The total number of inspected facilities in the EU report is 695 (≡ 100%). Of these, 613 facilities (≡ 88%) were found to be in compliance at one time or the other.

For 30 facilities (≡ 4%), a decision was pending (*e.g.*, re-inspection, answer to the inspection report, etc.), and 8 facilities (≡ 1%) were not in compliance. No clear status was shown in the list for the rest of the inspected facilities (44, or 6%).

During the inspection of some industry-owned facilities, the authorities extended the inspections into other fields, with special emphasis on:

France:	metrology, ISO 9000, EN 45000 (accreditation)
Germany:	working safety
Great Britain:	worker's health
Spain:	ISO 9000, EN 45000 (accreditation)

Moreover, the interpretation of some issues of the GLP principles differs considerably among the countries: *e.g.*, in the Netherlands, test facilities with 2 persons (Head and Study Director; QAU is contracted) have been certified; in Germany, a test facility has to consist of at least 8 persons: Head, Study Director, QAU, Archivist, and the corresponding deputy for each one of these functions.

Inspection Results

Table IV shows the results of some inspections carried out by the corresponding national authorities between 1990 and 1997 with regard to laboratory and field trial test facilities in Germany (8 facilities), France (3 facilities), Great Britain (3 facilities) and the Netherlands (one facility).

Table IV. Findings 1990–1997

Area	DEU[1]	FRA[2]	GBR[3]	NLD[4]	Total (%)
Org. & Pers.	18	2	9	2	31 (10)
QA Program	8	1	2	—	11 (3)
Facilities	4	—	6	—	10 (3)
App. & Reag.	25	9	5	1	40 (12)
Test System	3	—	—	—	3 (1)
Test Substance	2	—	—	—	2 (1)
SOPs	66	6	22	7	101 (31)
Study Conduct	51	2	23	2	78 (24)
Final Report	14	—	2	—	16 (5)
Archives	16	—	2	—	18 (6)
Other	9	1	—	1	11 (3)
Total	216	21	71	13	321 (100)

[1] Germany: 8 facilities (laboratory studies and residue field trials)
[2] France: 3 facilities (laboratory studies and residue field trials)
[3] Great Britain: 3 facilities (laboratory studies and residue field trials)
[4] Netherlands: 1 facility (residue field trials)

It shows the distribution of findings among the ten GLP issues as described in the OECD Principles. Apparatus and Reagents, Standard Operating Procedures (SOPs), and Study Conduct and Documentation make up for 219 (≡ 68%) of a total of 321 findings. The area of SOPs (101 findings ≡ 31%) seems to be especially problematic.

Future Perspectives

Eight years after Directive 87/18 came into force no harmonization regarding GLP has been reached in the EU. There are Member States still getting organized and without properly trained inspectors. Also, differences in the monitoring organization and procedures, experience of the inspectors, and interpretation of the principles make it difficult to obtain a speedy solution to the situation. If a quick solution is found, it tends to be the most rigid.

On top of that, the newly revised OECD Principles seem to contain certain issues which might burden the test facilities with more work, without improving overall quality (*8*).

Literature Cited

1. *Council Directive 87/18/EEC*, Official Journal of the European Communities No. L 15, 17.1.87, p. 29.
2. *OECD Good Laboratory Practice in the Testing of Chemicals, Council Decision C(81)30 (Final)*, adopted 12 May 1981, 2 Rue André-Pascal, F-75775 Paris Cedex, 1982.
3. *Council Directive 88/320/EEC*, Official Journal of the European Communities No. L 145, 11.6.88, p. 35.
4. *Council Directive 90/18/EEC*, Official Journal of the European Communities No. L 11, 13.1.90, p. 37.
5. *Annual GLP Inspection Report 1996*, European Commission, Directorate-General III, III/B/4/DB, Rue de la Loi 200, B-1049 Brussels.
6. *GLP-Info Nr. 6*, GLP-Bundesstelle, Bundesinstitut fuer gesundheitlichen Verbraucherschutz und Veterinaermedizin, Postfach 330013, D-14191 Berlin, August 1997.
7. *GLP-Info Nr. 4*, GLP-Bundesstelle, Bundesinstitut fuer gesundheitlichen Verbraucherschutz und Veterinaermedizin, Postfach 330013, D-14191 Berlin, July 1996.
8. *OECD Principles of Good Laboratory Practice (as revised in 1997)*, OECD Environmental Health and Safety Publications, Environment Directorate, 2 Rue André-Pascal, F-75775 Paris Cedex, 1998.

Chapter 27

Registration in France: A Changing Scene

GLP Certification: 2 Plays in One

Dominique Ambrosi and Christine Touratier

CFPI AGRO Regulatory and Registration Department, P.O. Box 75, F-92230 Gennevilliers, France

The transcription of the European Directive 91/414/EEC into French law resulted in the need for a reorganization of the registration of agrochemicals. Beside the two existing bodies, *i.e.*, the Toxicological Commission and the Registration Committee, a third party called 'Structure Scientifique Mixte (SSM)' will pool the experts required for the evaluation of the Dossiers for the active ingredients at the European level and for the formulated products at the French level. All the safety studies making up these Dossiers have to be conducted in laboratories working under Good Laboratory Practice, which in France are certified mainly by the 'Comité Français d'Accréditation (COFRAC)' (for industrial chemicals and agrochemicals) and occasionally (mammalian toxicological studies) by the 'Agence Française du Médicament (AFM)' (for pharmaceuticals).

Since the first European treaty was signed in Rome, the goal has been to open the market on a free trade basis with equality between Member States. In this respect, Directive 91/414/EEC (*1*) was the basis for a uniformalization of the registration procedures for agrochemicals at the European level, introducing a positive list of active substances, which would prevail in all countries. The differences between these coming only with the registration of the formulated products which still have to be adapted to the local agricultural practices and climatological conditions. This resulted clearly in an overload of the capabilities and resources of the French bodies in charge of registration and to a reorganization of these structures.

With respect to Directive 91/414/EEC, the OECD Guidelines (*2*) and OECD GLP principles (*3*) are part of this uniformalization and are the basis of a mutual recognition of the certification of the Research Organizations and of the tests they are conducting.

Registration in France

Until recently, registration of pesticides in France was covered by a law from November 2, 1943, which was one of the first pesticide regulations in the world. It has been amended on several occasions since that time, but the principle has remained the same. A registration dossier was made of three different and complementary parts:

- The 'administrative dossier' includes all the nonscientific information, such as, the detailed composition of the product, the target pests and application rate, the commercial name, etc.
- The 'toxicological dossier' is made up of all the studies necessary to allow a clear assessment of the risk that could result from the use of the new chemical or product toward humans (operator as well as consumer), wildlife or the environment.
- The 'biological dossier' includes in-life results demonstrating the efficacy of the new product on target pests, together with its selective action to the crops.

The Old Organization. Under this previous organization, two groups were successively involved: The Toxicological Commission and the Registration Committee.

The Toxicological Commission. This Commission was composed of 65 members from all the Ministries involved (Agriculture, Environment, Health, Industry, and Finances) as well as research bodies (especially INRA = National Institute for Agronomic Research), corporate organizations, and the agrochemical industry. An independent Rapporteur member was designated among them to review the Toxicological Dossier containing all information on the active ingredient and formulated product for physico-chemistry, mammalian toxicology, ecotoxicology, environmental fate, operator exposure and residues.

Initially, this review of the Dossier took from 12 to 18 months until the Rapporteur member presented his conclusion to the Commission, which was proposing a label that included Maximum Residue Limits (MRLs) for all food crops covered by the application and Pre-Harvest Intervals (PHIs). PHIs define the minimum time to observe between treatment and harvest in order to minimize the residues in the plant fraction of concern.

The Registration Committee. This Committee, consisting of 25 members (from the same ministries and from INRA), was then studying the Biological Dossier that dealt with the agronomic part of the package and included trials on the efficacy against the pest(s) to be controlled and on the sensitivity of the crop to be protected. This review was presented generally within 2 months after the examination of the Toxicological Dossier by the Toxicological Commission and the Registration was generally granted within 6 weeks after the review of the Biological Dossier, making a total of less than 2 years from application to registration.

During the last 50 years, the amount of 'toxicological' information required for registering a new product and, even more, a new active ingredient has been expanding dramatically, and consequently, overloading the Rapporteur members, who very often were voluntary participants who had their own expert profession aside of the Commission. Also, the extension of GLP to all types of studies, more or less related to the safety of the product, contributed to the increasing thickness and complexity of the registration files. Furthermore, the Directive 91/414/EEC issued on August 19, 1991, (and transposed into French law on May 7, 1994) created a European registration for the active ingredients which have to be on a European positive list, called 'Annex I', before any Member State could register a product containing it. The review of the file on the active ingredient (called Annex II) is conducted by a Member State chosen by the applicant.

In parallel to this organization for the new active ingredients, the EU started a re-registration program for the *ca* 900 active ingredients already existing in Europe.

A first list of 90 existing active ingredients to be reregistered was published on December 15, 1992: France was designated as the Rapporteur Member State for 11 of them.

In addition, France has been chosen by applicants for 11 out of 38 new active ingredients (as of September 1997) to review the Annex II Dossiers. This has led to an increase in the time required by the Toxicological Commission to review the Toxicological Dossier, thus the two-year time frame was exceeded in 1996.

Finally, there was a need for permanent scientists who would stay in contact with their European colleagues for the harmonization of the evaluations and the general discussions in Brussels.

The old organization, therefore, was insufficient by far, and there was an important need for permanent scientists in order to cope with the continuously increasing workload that was created by the modern registration requirements and the European registration/re-registration system.

The New Organization. In order to solve these problems, a new organization was developed, which maintained the previous basic structure of the file in 3 parts, but re-organized the pattern to be followed by the file in order to improve speed and efficiency (see Figure 1).

This new organization was announced by the French Ministry of Agriculture in a public conference on October 17, 1996: The two existing decision bodies will stay, although the Toxicological Commission will be reduced in number. However, the Rapporteur will belong to a new group called SSM (Mixed Scientific Structure), 'Mixed' meaning that it is reporting to both the INRA and the Food Administration (DGAl). SSM, who will be based close to the scientific teams of INRA in Versailles, will be constituted of 5 scientific coordinating units in charge of physico-chemistry, mammalian toxicology, residues, environmental fate and ecotoxicology. These 5 units will be headed by a Director (presently, Mr. A. B. Delmas, an environmentalist of INRA, who is already a member of the Toxicological Commission). For residues, in particular, it must be remembered that France is the only country belonging to both Northern Europe and Southern Europe and thus needs a double number of trials. The expert responsible for each unit will, of course, be a member of the Toxicological

Figure 1. Registration in France : The New Organization

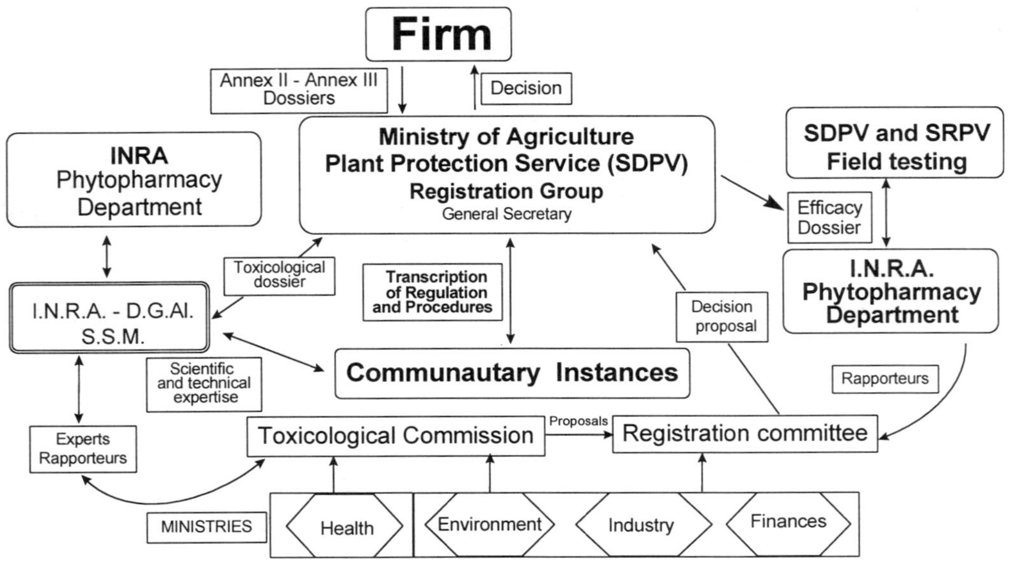

Commission. SSM will also be in charge of all the scientific contacts at the European level .

Under the new organization, the applicant should send the Registration Dossier to the Administrative Secretary of SDPV (Plant Protection Department of Ministry of Agriculture).

For a new active ingredient, for which France would be the Rapporteur member state, French Authorities will strictly apply the rules governing the new structure, with a preliminary check for completeness (including GLP compliance of the studies) of the Annex II Dossier within 6 months, followed by the review of the Toxicological Dossier. This review will result in a monograph, which will be presented and discussed at the European level. At this stage, there is no difference between any of the European countries.

For a new formulated product, the Annex III dossier, also made up mainly of GLP studies (in agreement with Directive 87/18/EEC) (*4*), conducted according to EU Guidelines (very close, if not identical to OECD Guidelines, in agreement with Directive 91/414/EEC), will then be sent to SSM who will review it in conjunction with European bodies and will transmit its conclusion to the Toxicological Commission. The Commission will issue a proposal for labeling. The Registration Committee will then review the Biological Dossier, now consisting of official (or officially recognized) trials conducted in agreement with Directive 93/71/EEC (*5*), which acknowledged the difficulty of implementing GLP in efficacy field trials and defined what we could call GEP (Good Experimental Practice). The Committee will issue a decision proposal for registering (or not registering) the pesticide to the Administrative Secretary of SDPV. This group, in coordination with Brussels, will make a decision and inform the applicant of the regulatory decision.

Registration of pesticides in France is, therefore, becoming part of the European registration system, and this is clearly in agreement with the ultimate goal of equality between Member States. In this respect, GLP also takes part in this fair competition at the European level in setting common standards to be adopted for the studies conducted for a registration dossier; thus, we are developing uniformity between Research Organizations, between applicants and even between registration authorities!

GLP Certification

All GLP regulations in France refer to the OECD principles as described in the OECD Council decision of May 12, 1981 [C(81)30(Final)]. However the recognition of these principles differ according to the field to be regulated: Chemical products (including pesticides) or pharmaceuticals.

Chemical Products. At the European level, all chemicals are regulated under the Directive 67/548/EEC (*6*) for the labeling and classification of 'dangerous substances'. However, this Directive clearly excluded from its frame several classes of chemicals, particularly, pesticides and pharmaceuticals. It also only concerns active substances and not their preparations.

This very famous, important, basic Directive is regularly amended by new Directives 'adapting 67/548/EEC to technical progress'. The first of these adaptations was the Directive 76/907/EEC (*7*) which reintroduced the pesticidal active ingredients into the frame of the Directive 67/548/EEC.

For GLP, the Directive 87/18/EEC asks the Member States to require and control the application of OECD GLP principles to all trials conducted according to Directive 67/548/EEC (therefore, to most chemicals, including pesticides, but still excluding pharmaceuticals) and to all facilities conducting such trials. These GLP principles were definitively adopted by Directive 89/569/EEC (*8*), at about the same time when OECD published their decision-recommendation on October 2, 1989, after the final agreement was reached in Paris in 1988, in the document C(89)87 (revised in 1995) (*9*).

This was completed by another Directive 88/320/EEC (*10*) describing in more detail the inspection and the control of GLP principles in the laboratories claiming this reference for their studies. This Directive was finally completed by Directive 90/18/EEC (*11*), which included as Annexes the corresponding full documents from OECD.

At the French level, the EEC Directives were transposed into French law under the Decree 90.206 (*12*). This decree included the OECD GLP principles in an Annex. It included how the inspection was to be conducted and how GLPs were to be implemented. These responsibilities were given to the already existing GIPC (Interministerial Group for Chemical Products), who delivered the certification after the inspection conducted by 'Réseau National d'Essais (RNE = National Network of Trials)'. Today's organization is the same, except that RNE was replaced in June 1994 by COFRAC (French Committee for Accreditation) for the inspections of the facilities and of the studies (and particularly by 'Section Essais du COFRAC'-37 rue de Lyon-75012 Paris, France).

The laboratory must first apply for certification, then fill in a form including information on the facility and a commitment to facilitate the inspections. This commitment is confirmed by the next step, which is an auto-evaluation of the adherence to GLP principles. This document prepares for the first inspection.

The first inspection will (or will not) result in a certificate that is valid for 1 year. The first 'regular' inspection occurs after 1 year has elapsed. These regular inspections will then occur every other year and will always include the audit of a study that is either already completed or ongoing. They will renew the certification for 2 years. During the year between the regular inspections, a questionnaire will have to be completed that shows the improvements made following last inspection, particularly for any deviations reported.

GLP certification is granted for a definite number of 'fields of expertise' which are listed here :

1) Physico-chemistry
2) Mammalian toxicity
3) Mutagenicity
4) Ecotoxicity (aquatic and terrestrial)
5) Environmental fate (water, soil, air) / Bioconcentration

6) Residues
7) Mesocosms and natural ecosystems
8) Analytical and clinical chemistry
9) Others

The cost of the first inspection is about $6,000 USD. Then, each year, there is an annual fee based on the number of technical units certified or accredited ($700 USD plus $240 USD per unit), and in addition, every other year, there is a periodic inspection, the cost of which is based on the number of inspectors involved, the duration of the inspection, and the expenses incurred by the inspectors. This makes the cost around $3,000 USD/year after the first inspection.

As of July 15, 1997, 47 contract laboratories have been certified (see Table I). The internal labs of major companies, like Agrevo, Ciba-Geigy (now Novartis), Cyanamid, Du Pont, Elf Atochem, FMC France, Hoechst, L'Oréal, Rhone-Poulenc, Roquette Frères, Roussel Uclaf, Sandoz (now also Novartis) and Sopra (Zeneca's subsidiary), were excluded from the Table.

Most often, at the same time as the certification, the inspection will also check for accreditation for the same category of trials or analyses, according to EN45001 (*13*), with particular emphasis on repeatability and reproducibility of the trials or analyses. For pesticides, this technical accreditation can participate in two programs : program N°95 devoted to the analysis of agrochemical products and program N°99.2 devoted to the analysis of pesticide residues in animals, feed and food.

Pharmaceuticals. Soon after OECD proposed their first version of GLP principles in 1981, the French Ministry of Health required in an instruction of May 31, 1983 (Decree of January 20, 1986) (*14*), that all 'experimental toxicology' tests, now called 'non-clinical' trials, and the safety tests, be conducted according to GLP principles, and this document described these principles in detail. An instruction of September 3, 1984 (Decree of January 20, 1986) (*15*) described how to conduct GLP inspections. The requirement for GLP was confirmed by the Directives 87/19/EEC (*16*) and 87/20/EEC (*17*). The list of the tests involved has been updated by a decision of December 9, 1996, which lists all topics that must appear in the registration file for a new pharmaceutical.

Since 1993, everything related to pharmaceuticals is dealt with by the AFM (French Agency for Medicines, 143-147 Boulevard Anatole France, 93285 Saint-Denis Cédex, France). In particular, GLP compliance is controlled by AFM inspectors. A program of joint inspections AFM/COFRAC for laboratories (only 4 at this time) working on all kinds of chemicals is currently being studied.

The principles of certification are about the same as for non-pharmaceutical compounds, and there is also a regular inspection every two years. As of February 1997, 32 laboratories were GLP certified by AFM to conduct preclinical and safety trials on pharmaceuticals. It must be noted that for the mammalian toxicological trials, studies conducted by contract laboratories certified by AFM also are accepted by the French Authorities in charge of the registration of pesticides (although there is no official recognition yet).

Table I. List of the Independent Laboratories GLP Certified by COFRAC.

Laboratory	Inspection	Category (OECD)*
ADME-Bioanalyses (06250 Mougins)	Regular (June 1996)	6
ANADIAG (67500 Haguenau)	Additional (June 1996)	1-6-8
Biotek Agriculture (10260 Fouchères)	Regular (May 1995)	6 (field)
Biovac (49071 Beaucouzé)	First (April 1996)	9 (animal experiment)
CERAAF (41000 Blois)	Regular (June 1996)	6
CIT (27005 Evreux)	Regular (April 1997)	1-2-3-4-6-8
CNEVA (35133 Javene)	Regular (February 1997)	3
CoopAgri Bretagne (29206 Landernau)	Regular (Nov. 1996)	6-8-9 (sensory analysis)
Laboratoire Crépin (76178 Rouen)	Regular (February 1995)	1-6-8
Défitraces (69126 Brindas)	Regular (January 1997)	6
European Ag. S. (69007 Lyon)	Regular (April 1996)	4-5-6-7-9(field-operator exposure)
EVIC CEBA (33290 Blanquefort)	Regular (Dec. 1994)	2
Fac.Pharmacie (MICRAAM 13385 Marseille)	Regular (April 1996)	9 (microbiology)
IFBM (54500 Vandoeuvre)	First (January 1997)	1-6-9 (mini-brewery)
INERIS (60550 Verneuil en Halatte)	Regular (many inspections between May 94 and May 96)	1-2-4-5-7-9 (pyrotechnic)
INRA GRAPPA (84914 Avignon)	First (January 1996)	6
Laboratoire LARA (31300 Toulouse)	Regular (March 1996)	6
LECMHA (69300 Clermont-Ferrand)	Regular (Nov. 1995)	9 (microbiology)
PrestAgro (38313 Bourgoin-Jallieu)	Regular (March 1997)	6 (field)
Société Promovert (64121 Serres-Castet)	Regular (January 1997)	6 (field)-7(field)
SEPC (69490 Sarcey)	Regular (October 1995)	1-4
Lab. Simon France (Wolff Group, 92110 Clichy)	Regular (Mars 1996)	1-8-9 (microbiological stability)
Stage (37100 Tours)	First (December 1996)	5-6
Staphyt (62860 Inchy en Artois)	Regular (January 1997)	6 (field)-7(field)
Société Viti R&D (34400 Villetelle)	Regular (April 1996) First (April 1996)	9 (micro-processing) 6 (field)
Laboratoires Wolff (92110 Clichy)	Regular (May 1996)	6

**1) Physico-chemistry 2) Mammalian toxicity 3) Mutagenicity 4) Ecotoxicity (aquatic and terrestrial) 5) Environmental fate/Bioconcentration 6) Residue analysis 7) Mesocosms and natural ecosystems 8) Analytical and clinical chemistry 9) Others*

Conclusion

In conclusion, it appears that, due to the pressure and the burden resulting from the European registration organization (but also due to the ever present willingness of France to participate in the building of Europe and, therefore, to harmonize regulations at the EU level), the registration organization for pesticides and the GLP certification system in France are becoming closer and closer to the systems prevailing in the two other main European countries, *i.e.*, the United-Kingdom, with its Pesticide Safety Directorate (PSD), and Germany, with its Biologische Bundesanstalt für Land- und Forstwirtschaft (BBA).

Literature Cited

1. Council Directive 91/414/EEC of 15 July 1991, Official Journal of the European Communities No. L 230, 19.08.91, p. 1.
2. OECD Guidelines for Testing of Chemicals, 1993, Head of Publications Service. OECD, 2 rue Andre-Pascal, 75775 Paris Cedex 16, France.
3. OECD GLP Principles C(81)30(Final) (adopted 12 May 1981).
4. Council Directive 87/18/EEC of 18 December 1986, Official Journal of the European Communities No. L 15, 17.01.87, p. 29.
5. Commission Directive 93/71/EEC of 27 July 1993, Official Journal of the European Communities No. L 221, 31.08.93, p. 27.
6. Council Directive 67/548/EEC of 27 June 1967, Official Journal of the European Communities No. 196, 16.08.67, p. 1.
7. Directive 76/907/EEC of 14 July 1976, Official Journal of the European Communities No. L360, 30.12.76, p. 1.
8. Council Directive 89/569/EEC of 28 July 1989, Official Journal of the European Communities No. L 315, 28.10.89, p. 1.
9. OECD Decision-Recommendation C(89)87 (adopted 2 October 1989), revised in C(95)8(Final) of 09.03.95.
10. Council Directive 88/320/EEC of 9 June 1988, Official Journal of the European Communities No. L 145, 11.06.88, p. 35.
11. Commission Directive 90/18/EEC of 18 December 1989, Official Journal of the European Communities No. L 011, 13.01.90, p. 37.
12. Decree No. 90-206 of 7 March 1990, Journal Officiel de la République Française, 9.03.90, p. 2891.
13. EN45001, 1989, General criteria for the operation of testing laboratories. European Committee for Standardization, rue de Stassart 36, 10050 Bruxelles, Belgium.
14. Decree of 20 January 1986 (Instruction of 31.05.83), Journal Officiel de la République Française,16.02.86, p. 2647.
15. Decree of 20 January 1986 (Instruction of 03.09.84), Journal Officiel de la République Française,16.02.86, p. 2647.
16. Directive 87/19/EEC of 22 December 1986, Official Journal of the European Communities No. L 15, 17.01.87, p. 31.
17. Directive 87/20/EEC of 22 December 1986, Official Journal of the European Communities No. L 15, 17.01.87, p. 34.

Chapter 28

GLP Considerations on the Road to Mutual Acceptability: A Swiss Perspective

Iris R. Wüthrich

Quality Assurance Unit, RCC Umweltchemie AG, Zelgliweg i, CH-4452 Itingen, Switzerland

Thanks to several agreements for the mutual acceptance of GLP between Switzerland and the OECD, U.S. EPA and FDA, as well as with different Japanese ministries, the Swiss GLP is accepted worldwide. Responsibility for the GLP compliance procedures in Switzerland lies with three authorities, the Federal Office of Public Health (BAG); the Federal Office of Environment, Forests and Landscape (BUWAL); and the International Office for the Control of Medicaments (IKS). The Swiss Guidelines, "Good Laboratory Practice in Switzerland, Procedures and Principles", March 1986, are based on the "Decision of the OECD Principles of Good Laboratory Practice," Paris, France, 1981. Besides GLP compliance, special reporting formats are required by different registration authorities, e.g., U.S. EPA PR Notice 86-5 or the German BBA. This can lead to complications since GLP allows only one final report.

The GLP Regulations in Switzerland were promoted to accommodate the large number of chemical and pharmaceutical companies and contract research organizations whose studies are being used for registration purposes worldwide. Compliant laboratories are inspected by the authorities every two years.

Swiss GLP studies are accepted by the submission authorities all over the world; for instance, in the different EU countries, the USA, Japan and other Far Eastern countries.

Regulatory Basis for GLP In Switzerland

The Swiss GLP Monitoring System operates in the following manner:

Firstly, the testing facility intending to perform GLP studies has to apply for an authority inspection;

Secondly, the facility will be inspected by representatives of the concerned authorities; and

Finally, after having successfully passed an inspection, the facility receives:

- An inspection report describing the observations, findings and deficiencies.
- An official decision from the monitoring authority concerning the GLP compliance status.

For purposes of information, the official decision also is included in two further documents, namely:

- As a Statement of GLP Compliance, issued by the Department of the Interior. This certificate-like document is given to the testing facility for use as a submission document in countries like Germany.
- As an entry in the annual report of the GLP authorities, which contains an inventory of all testing facilities inspected. This document is exchanged for information between GLP monitoring authorities of the OECD member countries.

Usually after two years, but not later than three years, the inspection is repeated and the procedure described above begins anew.

The Swiss GLP Guidelines. The present Swiss guidelines, "Good Laboratory Practice in Switzerland, Procedures and Principles" (*1*), March 1986, are based on the OECD Principles of Good Laboratory Practice of May 12, 1981 (*2*). This also includes the recommendation of the OECD Council concerning the Mutual Recognition of Compliance with GLP of 1983.

Since the early eighties, however, times have changed a great deal in the Swiss GLP world: The requirements of the authorities have increased and GLP has expanded into new areas. Multi-site studies, Principal Investigators, electronic data processing, etc., are some examples. Therefore, several OECD Consensus Documents have been issued in the last few years describing how GLP should be applied.

Some excerpts from several Consensus Documents have been included in the newly revised (1997) and published (January 1998) OECD GLP guidelines. As Switzerland is also an OECD Member State, the GLP changes described in the different Consensus Documents are normally included in daily GLP work as soon as the respective papers have been issued. In addition, once the 1997 OECD GLP guidelines (*3*) are officially translated into German, the Swiss GLP guidelines (1986) will be revised accordingly.

In Switzerland four official languages are recognized. The GLP guidelines are available in three of these languages: German, French and Italian, as well as in the internationally acceptable language, English.

The Swiss GLP Authorities. The responsibility for monitoring GLP compliance in Switzerland is assigned to the Federal Office of Public Health (BAG), the Federal

Office of Environment, Forests and Landscape (BUWAL), and the Intercantonal Office for the Control of Medicaments (IKS), (Figure 1 and Table I).

Table I. Delineation of Responsibilities

Types of Studies	Types of Chemicals		
	Pharmaceuticals	Agrochemicals	Industrial Chemicals
Physical/Chemical	IKS	BAG or BUWAL	BAG or BUWAL
Toxicological	IKS	BAG	BAG
Ecotoxicological or Environmental	BUWAL	BUWAL	BUWAL

The GLP Inspection Procedure. The GLP authority inspection process begins when a testing facility requests an inspection by writing to the responsible authorities. With a view towards international acceptance, a frequency of routine inspections of approximately every two years is recommended.

Small testing facilities are subsequently inspected by one of the three authorities (according to Table I). The duration of the inspection varies depending on the size of the testing facility and the findings of earlier inspections. Usually, the inspection team consists of two inspectors.

Larger testing facilities with activities in several fields may be inspected by more than one authority. Upon request of the testing facility inspections are performed in a coordinated way by the three responsible authorities. Test facilities that perform tests for different types of chemicals within the same laboratories may apply for a joint inspection of all authorities involved.

The inspections are conducted in accordance with the OECD Series on Principles of Good Laboratory Practice and Compliance Monitoring: Guidance and Consensus Documents (No. 2-10) (*4*). A routine inspection of our company, RCC, with approximately 400 staff, requires 30 man-days of inspection. In practice, this means that six inspectors spend one week at RCC.

After the inspection, the testing facility receives the authority's inspection report containing, among other things, the result of the inspection, specific findings and conclusions concerning compliance or noncompliance. Inspection reports are presented to the test facility in draft form for comments of inconsistencies. The test facility responds in writing, thus confirming the findings of the inspectors and where required, explaining what corrective actions will be taken to avoid future deficiencies. If the deficiencies found were only of minor importance and did not interfere with the integrity of data, and if the authorities are satisfied with the corrective actions promised by the testing facility in writing, the authorities issue an official decision confirming GLP compliance. Should, however, major deviations

Figure 1. The Agencies Responsible for GLP in Switzerland

THE SWISS GLP AUTHORITIES

INTERKANTONALE KONTROLLSTELLE FÜR HEILMITTEL
OFFICE INTERCANTONAL DE CONTRÔLE DES MÉDICAMENTS
UFFICIO INTERCANTONALE DI CONTROLLO DEI MEDICAMENTI

BUWAL, Bundesamt für Umwelt, Wald und Landschaft
OFEFP, Office fédéral de l'environnement, des forêts et du paysage
UFAFP, Ufficio federale dell'ambiente, delle foreste e del paesaggio
FOEFL, Federal Office of Environment, Forests and Landscape

Bundesamt für Gesundheitswesen
Office fédéral de la santé publique
Ufficio federale della sanità pubblica
Swiss Federal Office of Public Health

from GLP be detected or only insufficient corrective actions be proposed, a negative decision may be taken. The testing facility will be mentioned in the annual report as not confirmed, and no Statement of GLP Compliance will be issued by the Department of the Interior.

At present, the Swiss GLP compliance-monitoring program encompasses 70 testing facilities.

Swiss GLP Compliance Statement. If the inspection confirms that the principles of GLP are respected and followed, the Federal Department of the Interior will issue a GLP Statement of Compliance (usually written in English) containing the inspection dates and the names of the certified testing facilities (Figure 2).

The areas of expertise for which the facilities are GLP certified are mentioned in the authorities' inspection reports and on the GLP Compliance Statement.

International Acceptance of Studies Performed in Switzerland

As a Swiss contract laboratory, RCC often finds itself in the situation where, at the special request of our sponsors, we have to take into account not only the Swiss GLP guideline, to which we are bound, but also to consider other national regulations, such as, the German "Chemikaliengesetz" or the American EPA (*5*) or FDA GLP Regulations.

Based on the OECD decision on Mutual Acceptance of Data and due to additional bilateral agreements with other OECD member countries, the Swiss GLP Guidelines are recognized widely around the world, and studies performed according to them are accepted.

In Europe, a bilateral agreement exists today between Germany (*6*) and Switzerland. Negotiations for a Mutual Recognition Agreement are currently ongoing with the Commission of the European Union.

Further agreements have been signed with several Japanese ministries, namely, the Ministry of Agriculture, Forestry and Fisheries (MAFF) (*7*) in 1993 and the Ministry of Health and Welfare (MHW) and the Ministry of International Trade and Industry (MITI) (*8*) in 1994.

In addition, a memorandum of understanding was signed between the Swiss authority and the American FDA in 1985 (*9*), and with the EPA in 1988 (*10*).

An extract of the memorandum of understanding (MOU) between the U.S. EPA and the Swiss authority reads:

"Although not identical, the GLP principles of both parties are comparable and, therefore, mutually acceptable. They are adequate to foster the collection of quality data and satisfy the Principles of Good Laboratory Practice recommended by the OECD."

The respective agencies of both countries assess adherence to the principles of good laboratory practice through the conduct of periodic inspections by a trained government inspectorate.

Figure 2. Swiss GLP Compliance Statement

EIDGENÖSSISCHES DEPARTEMENT DES INNERN
DÉPARTEMENT FÉDÉRAL DE L'INTÉRIEUR
DIPARTIMENTO FEDERALE DELL'INTERNO

GLP Compliance Statement

It is hereby certified that

on	February 12-16, 1996 February 19-23, 1996 June 14, 1996
the testing facilities of	**RCC Holding Company Ltd** **4414 Füllinsdorf** **Switzerland**

were inspected by the Federal Office of Public Health, the Federal Office of Environment, Forests and Landscape and the Intercantonal Office for the Control of Medicaments with respect to the compliance with the Swiss GLP Principles. The inspection was performed in agreement with the OECD Guidelines for National GLP Inspections and Audits and comprised the following testing facilities:

- RCC Research and Consulting Company Ltd, Itingen
- RCC Umweltchemie AG, Itingen
- RCC Pharmanalytics Ltd, Itingen
- BRL Biological Research Laboratories Ltd/Microbiology, Füllinsdorf

It was found that the aforementioned testing facilities were operating in compliance with the Swiss Principles of Good Laboratory Practice (Good Laboratory Practice [GLP] in Switzerland, Procedures and Principles, March 1986) at the time they were inspected.

FEDERAL DEPARTMENT OF THE INTERIOR

Bern, July 9, 1996

Ruth Dreifuss
Federal Councillor

The inspection programs, which are mutually acceptable, permit assessment of current laboratory and field operations, as well as the audit of final reports of selected studies.

Switzerland, for example, was recently (1996) visited by the U.S. FDA. The last inspection performed at one of RCC's testing facilities by the FDA-equivalent Swiss authority was attended and observed by Mr. Stan Woollen, Director, Division of Compliance Policy, FDA. As a result of this inspection, RCC received a written confirmation from the Swiss authority. The EPA has announced that it plans to inspect the Swiss authorities some time in 1998.

GLP Regulations Versus Different Reporting Requirements for Registration

The reporting requirements from the different registration authorities, e.g., the German BBA, the EU and the American EPA PR Notice 86-5 (*11*), may have an impact on the GLP compliance of a finalized report. The first example relates to the general report format. The differences become apparent within the first few pages.

A report written for submission for a German registration of an agrochemical requires, on the second page, a copy of the authority's GLP Compliance Statement of the testing facility where the study was performed. A report for EPA submission - on the other hand - doesn't need this GLP Compliance Statement, but it does require, the following (*12*):

- the statement of data (no) confidentiality claims (page 2)
- the GLP compliance statement (page 3)
- certification by the applicant that the report is an unaltered copy as received from the testing facility (page 4)
- the QAU statement (page 5)

If a pharmaceutical/chemical company wishes to submit his EU report to the United States EPA, the administrative pages of this document must be modified according to the submission requirements of the EPA PR Notice 86-5 and the testing guidelines, as described above.

Statements of Compliance: EU versus EPA. The second example concerns the GLP compliance statement. For EU submissions, the Statement of Compliance is only required to be signed by the Study Director. The GLP Statement of Compliance for EPA submissions, however, requires the sponsor to sign the Statement of Compliance in addition to the Study Director, as both bear joint responsibility for the performance of the study in conformity with GLP. A third signature has to be given by the person submitting the study to the authorities.

Concerning the Statement of Compliance, RCC was once faced with the following problem:

A European agrochemical company was taken over by an American company. Consequently, the American company became the owner of the studies previously performed and, naturally, wanted to use them in the U.S. for submission.

The studies performed almost a decade ago were intended for European submission only. The respective reports contained the European version of the Statement of Compliance, signed only by the Study Director and not, as required by 40 CFR 160, with the Study Sponsor's and Submitter's signatures in addition.

Occasionally, reports which were finalized more than ten years ago, do not contain a Statement of Compliance. At that time, neither the OECD nor the Swiss or German GLP guidelines required one.

A representative of the American company, mentioned above, wanted to submit the study reports to the U.S. EPA authority. Not being familiar with the MOU between the U.S. and Switzerland, he asked RCC to add or to revise a Statement of Compliance - revised in the sense of a rewritten and newly signed statement.

How could the problem best be solved? If the Study Director was still employed by RCC, the only solution would be to modify the final report to comply with the EPA submission requirements. However, what would happen if the Study Director had left the testing facility? For the studies concerned, the Study Director must be changed by amendment to the report. That would mean, however, that a "new" Study Director would have to sign a statement for a study he does not know and was never involved in! Such an action would be a real GLP violation and, therefore, unacceptable.

A further point which must be taken into account is the Study Sponsor's signature. The original Sponsor Representative may have another position within the company or even works for a different organization.

Considering these facts, it would be impossible to revise the Statement of Compliance since this would violate the point of GLP. The situation was discussed with the Swiss GLP authority in order to find a solution. They referred to the existing MOU between EPA and Switzerland. In the scope of the mutual acceptability, the old European reports must be accepted as they are, without any revised Statement of Compliance. The American company was informed accordingly and they accepted our proposed procedure.

Conclusion

The objectives of the GLP regulations are similar in all countries, namely, the performance of scientifically sound and well documented studies. Thanks to mutual acceptance, duplication of studies can be avoided and, therefore, animals, materials and money can be saved! Hopefully, the authorities of all OECD member nations will, in future, accept studies in spite of some variations which may occur in the GLP regulations of these different countries.

Finally, it should be emphasized that good communication between all involved is essential in order to create a solid basis of confidence between the testing facility, the sponsor and the different authorities. This is a vital prerequisite for the success of any study.

Acknowledgments

Special thanks are due to Dr. H. A. Hosbach of the Federal Office of Environment, Forests and Landscape (BUWAL) and Dr. M. J. Mirbach of RCC Umweltchemie AG for their review and constructive comments, and to S. Pendley for preparing the text.

Literature Cited

1. Federal Department of the Interior. *Good Laboratory Practice (GLP) in Switzerland, Procedures and Principles;* March 1986.
2. *OECD Principles of Good Laboratory Practice*; May 12, 1981. Republished as document *OECD/GD (92) 32* in 1992.
3. OECD Principles of Good Laboratory Practice. *[C(97)186/Final];* as revised in 1997.
4. OECD Series on Principles of Good Laboratory Practice and Compliance Monitoring. *GLP Consensus Document No. 2-10;* 1995.
5. Environmental Protection Agency; 40 CFR Part 160. *Federal Insecticide, Fungicide and Rodenticide Act, Final Rule;* August 17, 1989.
6. Bundesministerium für Umwelt, Naturschutz und Reaktorsicherheit. *Bekanntmachung über Vereinbarungen zwischen der Bundesrepublik Deutschland einerseits und Österreich, der Schweiz und den Vereinigten Staaten von Amerika andererseits über die Gute Laborpraxis bei Arzneimitteln, Chemikalien und Pflanzenschutzmitteln;* 8. Februar 1989.
7. The Ministry of Agriculture, Forestry and Fisheries of the Japanese Government and the Federal Department for Economic Affairs of the Swiss Confederation. *Arrangement;* January 18, 1993.
8. Ministry of Health and Welfare and the Ministry of International Trade and Industry of the Japanese Government and the Federal Department for Economic Affairs of the Swiss Confederation. *Arrangement;* March 28, 1994.
9. The Federal Office for Foreign Economic Affairs, Federal Department of Public Economy of the Swiss Confederation and the Food and Drug Administration U.S. Department of Health and Human Services. *Memorandum of Understanding;* April 29, 1985.
10. The Federal Office for Foreign Economic Affairs, Federal Department of Public Economy of the Swiss Confederation and the United States Environmental Protection Agency. *Memorandum of Understanding;* June 22, 1988.
11. Akerman, J. W. *PR Notice 86-5, Notice to Producers, Formulators, Distributors and Registrants;* U.S. Environmental Protection Agency: Washington D. C., July 29, 1986.
12. Garner, W. Y., Barge, M. S., Ussary, J. P. *Good Laboratory Practice Standards; Applications for Field and Laboratory Studies;* American Chemical Society: Washington D. C., 1992.

Chapter 29

Ecotoxicology and Good Laboratory Practice: European Requirements and Practices

Nigel J. Dent

Country Consultancy, Copper Beeches, Gayton Road, Milton Malsor NN7 3AA, England

This chapter aims to cover the regulatory aspects of ecotoxicity and review the practicalities and problems while putting forward some solutions to carrying out ecotoxicity studies according to GLP.

This certainly is not a review of the regulatory requirements, although attempts have been made to give some clarification in the process of the generation of Directives and their associated guidelines while concentrating primarily on the practical aspects of conducting field studies and the common problems that are encountered by Study Directors and field workers.

EU Pesticide Legislation

Pesticides have been subject to a great deal of legislative controls in Europe ever since the introduction of **Directive 67/548/EEC** (*1*), which covers the classification and labeling of dangerous substances. A subsequent Directive 91/414/EEC (*2*), known colloquially as the "***Authorizations Directive,***" was intended to provide a more coherent and comprehensive system of regulating agricultural pesticides and, more importantly, to harmonize the national provisions found across the EU. Currently, this system could best be described as in disarray.

The stated aims for the Directive are:

1. **To remove trade barriers** in plant protection products by **harmonizing registration procedures** and member states' regulatory testing requirements.

2. **To ensure that risks** to human and animal health, the environment and ground water **take priority over plant protection.**

3. **To prevent the needless repetition of tests on animals.**

The aim of mutual recognition by national authorities for finished products, therefore, is inherent in the key aspects of the Directive which specifies:

* The information required to allow active substances and products to be assessed.
* Provisions for data sharing between interested parties.
* Provisions for periodic review of substances on the Community list.
* Provisions for exchange of information between member states.
* Provisions for mutual recognition of authorizations granted by member states.

Directive 91/414: This was the framework for achieving these points. Most of the necessary procedures and technical detail in evaluation criteria have been detailed by amending directives and other regulatory processes.

However, despite the existence of this framework Directive, **there is still a considerable scope for confusion** due to the need for frequent adaptation of policy and procedures regarding technical progress, which in turn leads to changes to many of the relevant directives and regulations.

Looking at the activities in Brussels and the rapid pace of agricultural development in this complex area, it would be nearly impossible to produce a guide to the legislation, as this would soon be out of date.

EU Decision Making Process and Legal Instruments

As you may be aware, there is a complex procedure for making policy decisions, and it is briefly the following:

1. **Commission Proposals** - The Commission is basically responsible to the European Parliament and is the overseer of the Treaty establishing the European Community. Its members are appointed by joint agreement among the governments and expected to act in the interests of the EU.

2. **Parliament Opinion** - European Parliament monitors the work of the Commission and can vote a motion of censure while having budgetary powers.

3. **Economic and Social Committee Opinion** - This Committee is an advisory body but must be consulted before any wide ranging decisions are taken.

4. **Council Debates and Decisions** - The Council administrators represent the government of the member states, and proposals are considered by these senior officials who are instructed by the Council.

Results from this decision making process give rise to legal instruments, such as:

1. **Directives** that impose standardization across the EU in specifying dates by which member countries must introduce the necessary national legislation and administrative systems to implement the Directive.

2. **Council Directives** which provide a legislative framework and define specific objectives.

3. **Regulations** that have the force of law and are binding in their entirety while applicable across the EU.

4. **Decisions** which are administrative acts and are binding on those whom it may concern, e.g., a member state, a firm, or an individual.

Rather than delve in detail into the key aspects of this complex matter, I thought it would be appropriate to indicate the burden placed on the industry by enumerating the following documents. As we go through this set of documents one can liken it to the familiar Christmas carol, "*The Twelve Days of Christmas.*"

1. The *basic rules* on pesticides are contained in *at least four council directives* and *two council regulations;*

2. *Guidance on classification and labeling* of pesticides, again, is primarily governed by *five council directives and one proposed council directive;*

3. *Existing active substances* as provided for by article eight (*2*) of directive 91/414/EEC are contained in *four commission regulations, three commission decisions, two regulations in preparation and three documents;*

4. *Data requirements* under directive 91/414 are covered by *eight directives;*

5. *Test guidelines* basically consist of *six council directives;* and

6. *Evaluation of plant protection products* are governed by <u>at least</u> *eight council directives.*

Would you believe, therefore, that to conduct any aspect of pesticide work to comply with current legislation, the subsequent experimental work has to pay attention to, and in many instance heed most fervently, **23 council directives, 2 council regulations, 3 commission decisions, 4 commission regulations, 8 community directives, 1 proposed directive, 2 regulations in progress and 3 documents?** *In all, 46 documents are necessary to be considered*!

The current European legislation for pesticides is, as has been stated, **Directive EU 91/414**, which was revised and issued *in March 1996*. Requirements for the active ingredients are now described in annexes 1 and 2 in the recent publication.

The non-target arthropod component of this directive resulted from a workshop held in Wageningen, Holland in 1995. This was the ESCORT workshop, which breaking down the acronym is European Standard Characteristics of Regulatory Testing. One must admit that when first reading the title of the workshop, it did have other inferences and wider connotations in our current day to day activities!

The workshop was funded by the EU and driven by regulators who wanted the experts to reach some kind of agreement. Some parts, in practice, did not work too well. For example, there is little point in testing predatory mites and pre-emergence herbicides when the two can never meet. It is likely, therefore, that there will be a revision and a **second Escort workshop** to unravel the impractical requirements.

As it stands, all products must be tested on two standard sensitive species as well as **two further species** relevant to the likely use of the product. Therefore, a cereal product might test the beetle, *Peocilus Cupreus,* and the spider, *Pardosa*, where an orchard product would look at predatory Heteroptera and maybe a lacewing. These Tier 1 laboratory tests are carried out on glass plate or sand and involve a single rate equivalent to the maximum proposed commercial rate. When multiple applications are envisaged then twice the maximum rate must be tested. Harmlessness in these tests results in no further requirement for testing. Harmfulness results in a label saying so and the need for further testing which is more realistic to try to evaluate the likely risk from field use. This may take the form of a semi-field or field study.

Rather than getting involved in more tongue twisting taxonomy or further interpretations of soporific legislation, the latter I will leave for those insomniacs, I would now turn to my key issue which is the practicality of implementing **another Directive, 88/320/EEC** (*3*) - yes, "the application of Good Laboratory Practice" in areas such as ecotoxicology.

Practicalities

What types of studies can we envisage?

1. Soil dissipation
2. Effect on beneficial arthropods.
3. Environmental fate and ground water studies.
4. Effect of veterinary products being "passed" through the host or recipient into the environment.
5. Combinations of these.

Where do these studies take place?

1. Any worldwide open air location irrespective of the weather, flora and fauna.
2. Usually at short notice.
3. Usually in the most remote location for monitors, QA and field operatives.

How are these studies conducted?

1. By means of a spray application.
2. By the use of a host, e.g., the cow.

When are these studies conducted?

1. At an appropriate time to coincide with the growing season.
2. When there is an abundance of the "recipient to retest".
3. Over several seasons.

Conclusions From These Practicalities

It is, therefore, of prime importance that **essential planning** is carried out to ensure the **availability of resources**, such as, staff, material, that as much **advance warning** as possible of weather conditions is sought and, of course, the **abundance of the "entomological beast"**.

As can be seen, these studies are much more difficult to conduct, than the routine Good Laboratory Practice study where the captive "animal," mouse or cow, can be relatively easily restrained and controlled.

It is unlike the human GCP patient who, although fickle and often non-compliant, is again quite controllable.

The GCPV aspect of conducting other "clinical field studies" suffers similar dilemmas, but usually with a good Co-Investigator, or Principle Investigator, the study aim can be satisfactorily achieved.

To conclude this section on practicalities, I would suggest to you that the now accepted Principle Investigator syndrome, detailed in the recently published New Good Laboratory Practices, has helped the ecotoxicology field situation with an analytical laboratory component and a remote site situation within a multi-site or multinational study.

Problems - The Ecotoxicology Top Ten

Weather. This is the number one problem and often brings the dilemma of "will it allow spraying; will it effect the host, the predator and will it allow growth; will it drown the pitfall traps; or will it allow the Quality Assurance person to attend in a fine environment for achieving quality assurance *and a sun tan."*

Availability of Staff. This is the number two problem. There are often too few in peak times and insufficient numbers available for multi-site studies, too many available out of season, or some not trained in the appropriate discipline resulting in a failure to comply with the sponsor's protocol.

Study Director or Principal Investigator - That is the Question. The third dilemma - will the Study Director who is an expert in the field situation be happy to control the analytical assays in the laboratory? Will the Study Director in the analytical laboratory comprehend spray applications, pitfall trap sampling, and D-VAC sampling, or will they just assume that this is something so alien that it has nothing to do with them? Thankfully, with the rewritten OECD GLPs, we now have *the shining knight* called the Principal Investigator, who will frequently undertake the field situation control while working under and with the analytical laboratory Study Director. Thus, both parties are familiar with each of their own tasks and, in working together, can ensure compliance with the Protocol, procedures and Good Laboratory Practice at each of their respective sites.

The Entomological or Environmental Subject. The number four problem - are they present, in sufficient numbers and readily identifiable? Can you obtain "a good soil core"? Are the aquatic beings present in sufficient numbers and of the right species?

Test Product. The fifth of the ecotoxicology top ten poses such problems as availability of the product, shipment of the product, suitable expiry dates and certificates of analysis, ability to travel across borders with minimum delays and absence of sitting on runways at fifty degrees, and availability of relevant safety data and appropriate handling details. If this is not enough, then we have to discuss and develop the plans on how to dispose of the test products and reference compounds, especially if one of these is a non-marketed product.

One of the biggest problems for contract research organizations at the moment is that the label on the bottle often does not have the same name as the test substance described in the protocol. In one instance, it was even worse in that the Certificate of Analysis had a third name. With mergers of companies, the prefixes, or even whole identifiers, are being altered. In these instances, the Study Director is usually right to wait and sort out the test substance names, Certificates of Analysis, etc. If the only good weather window passes in the meantime, then that is the Sponsor's problem. It takes a pretty bold Study Director to hold up the train and do this, but then such confidence should be a requirement for Study Directorship anyway.

Quality Assurance. The problems here are a little simpler but can be complex. Is the QA person aware of ecotoxicology, is it a question of "have checklist can travel," or are they present in sufficient numbers and able to arrive on time at the study site? Is it possible that they can remain at the study site in the presence of bad weather to wait for the sun to shine and the sprayer to start?

Are the people experienced in carrying out audits of field study work and are they aware of the complexities of tank mixes, safe handling of product and where to stand to avoid being hit by a tractor and boom sprayer while avoiding the drift in a slightly windy condition? Finally, is it possible to be aware of all the requirements of these activities and to be able to assure the quality in between seasons?

The Report. Arriving at number seven of the top ten, the key problem is, is there the time required to produce the report in between the growing and spraying seasons. Is there Quality Assurance availability to audit the report and are they aware of the complexity and vast numbers of data points frequently required to be reviewed? If a Contract Research Organization (CRO) is involved, will the report meet the sponsor's time frame? Are these reports to be composite, i.e., will statistics be separate or will the field study part be separate from the analytical report, and who brings everything together in the appropriate time frame with the right summary and conclusion?

The Surrounding Environment. Of all of the points dealt with so far, this often is the most hostile and unpredictable. What other crops are there adjacent to the test area? Are there rivers and streams close by which could pose a problem for contamination for ground water? In the conduct of a field study, where several spray applications are desirable, will the farmer be aware of any silage cutting requirements and disposal, is this activity to take place and is there collaboration between operative of the ecotoxicology study and the farmer? What about the problems of release of animals onto the test site, the release of "environmentally friendly and organic fertilizer spreading", the rodents in the weather station, and the cows putting their feet in the areas where leaf bags have been disposed of?

Couple this with sheep running away with the tops of pitfall traps attached to their legs and, most importantly, the general public walking through the area removing marker canes, posts and generally interfering in what looks to be a very interesting area but should be totally restricted from their imposition. If this is not acceptable, then what happens when all of these activities are undertaken by a CRO?

The Equipment. As we get closer to the top of the top ten, we can now see that particular *"laws"*, which are one of the key elements certainly, will always be brought into play in this particular area, along with our colleague Murphy. Here we have the reliable if not temperamental D-VAC suction equipment, motorized tractor units, boom sprayers with a tendency to have nozzles blocking at the most inopportune moments, and back-pack sprayers that run out of compressed air or petrol when the last one liter of spray application remains to be dealt with. The particular lie of the land would allow the complete use of all of the tank mix, apart from the fact that the tractor driver has started in the wrong circle, and thus, the slope of the land does not allow the last 1 liter of spray application to be sucked up into the boom sprayer because in actual fact the solution lies away from the outlet in the bottom of the tank. Despite the best laid plans, the well calibrated and maintained equipment, looked at and prepared during the winter months, has nothing to prevent it from breaking down at the most awkward moment; and, therefore, the question arises, " *what about back-up equipment and site calibration ?"*

The Regulatory Environment. If it is not sufficient to deal with items one through nine that pose problems for carrying out an ecotoxicology study, then reviewing, revising and reading the 46 other documents before one starts is enough to put off the most hardy ecotoxicologist.

Solutions

Weather. The only real solution to this aspect is to buy a crystal ball, befriend a weather forecaster or water diviner or choose countries with as "standard a weather pattern" as possible.

However, you will find that experienced field workers become fairly advanced weather forecasters. Field staff interpret the radio shipping forecasts daily during the field season and draw isobar maps, buy the papers for their analyses and ring up the coast guard and similar persons before spraying to get as much information as possible.

Basically, there is little we can do in this particular situation.

Staff. Recruit and train temporary staff to allow for the peaks and troughs. Have on hand contract staff and build a good and "secure" team and work to a manageable trial volume level.

Within a company, do not be bullied by the sales force and marketing people - a good study is "a well designed, planned and timely executed study".

REMEMBER : "A poor design cannot benefit from accurate QA"

Study Director or Principal Investigator. With the newly developed OECD GLPs, the specter of the Principal Investigator has now become a reality.

The concept of the lone Study Director, based in the field, being responsible for the analysis *back at the ranch,* often carried out miles, or even continents away, can now become a recognized compliant item. Therefore, the Principal Investigator, who is primarily A Study Director *with less responsibilities,* can assist and overcome the problems of one study plan and the possible requirement of two Study Directors which is totally unacceptable to the Regulatory Inspectorate.

Surely this concept can only help industry in producing a more controlled study, with the key site based individual trained in a responsible area and for those elements they know best.

The Entomological or Environmental Subjects. Here the key watch words are ***look, select, check and recheck***. ***"If it is not available, there is little you can do."***

It is the Study Director and the Principal Investigator's responsibility to ensure the study design takes into account **the time of year, the availability of the subject** in the area and, like performing a good GCP study, **site selection before initiation** should be the key practical approach before embarking on the spraying.

Now we must turn to the availability of the species under review and test. Take the humble pitfall trap or the less than quiet D-VAC suction apparatus, resonating around the hillside. A typical "day's catch" from a pitfall trap when one of my clients conducted a three-year study was 300,000 specimens in one season. A single sample could contain up to 2,000 individual arthropods, often with very high numbers of the smallest ones. Was he lucky ?? **Yes, from the effect of the product and so many "bugs"** but **no, for the identification, preservation and recording aspects**. These are the typical catches of the day which enable a very efficient study to be carried out, but the key solution is "do you have the time and expertise to deal with this volume and variance"? Training and experience, therefore, is one of the keys to the solution.

"***It is the early toxicologist that catches the worms***" - a key parameter in some studies. Written into the Protocol is "by immersing the site in 0.2% formalin, we will wait for the worms to rise and then collect and identify".

What if the ground is so dry they have migrated to the center of the earth, and what if it's too wet and they have migrated, waiting for a dryer climate? Again we bow to the uncontrollable - the weather. However, one of the solutions is planning, but again with the weather this is not always the key solution. The time frame, however, is a possible benefit in dealing with this type of subject.

How often have we located a site, predetermined the availability of the "subject" only to find on the day of spraying they have migrated into another field? Perhaps, as has often happened, the converse applies. Another species is abundant and, therefore, extra evaluation can be made - a bonus to the ecotoxicologist and the sponsor. **Again pre-site evaluation is the key to this solution.**

Test Product. As we know, problems often occur where the product is to be dispatched directly from the sponsor to the site. Prevention of receipt by border controls and lack of availability of the correct amount or the correct batch are frequently occurring problems. Solutions to this can be to identify in advance potential bottle necks or problem areas. Prepare as far in advance as possible the paperwork, liaise with the regulatory bodies and the customs officials, and check with the production department.

To avoid some of the other problems identified earlier, we should notify the manufacturer, sponsor or local office as far as possible in advance. If a wholesaler is involved and the supply is to be purchased locally, checks on availability are very important. We must ensure that there is adequate quantity, good quality, same or similar batches, and that these can be made available on time.

The preplanning stage must ensure that the products are able to be delivered to the correct location and to be stored safely on the site, that there are good mixing instructions, that handling precautions are detailed, and that appropriate safety profiles are available. Discussions and plans for disposal in a safe and acceptable manner are next in order of priority. **Therefore, the solution to all these problems is, once again, good design, communication, liaison and well-planned activities.**

The final aspect in this area is the best guess amount per hectare, and to notify the product supplier of the amount of chemical needed plus 10% several weeks before spraying is to take place.

However, determining the quantity of test substance to order is a detailed calculation based on the area to be treated, spray tank volume, dead volume, etc., combined with an increase, up to 50% extra, to allow for unforeseen circumstances.

Quality Assurance. In terms of solving the problems that already have been identified, there is very little that can be achieved here unless this function can be contracted out. Often, Quality Assurance is overloaded, being on the critical pathway with few extra resources. One of the key areas may be to devote one member of the team solely to ecotoxicology studies and train accordingly.

Notification to, or identification of, sites that are required for audit well in advance will allow sufficient time to be built into the audit calendar. Knowing the fickle state of the weather and other aspects governing these studies, a certain amount of slippage either way should be built into their schedule by the Quality Assurance unit.

This, in general terms, could be quoted as "weather time" so that if they arrive on site and conditions prevent activities they can move to an adjacent tavern, carry out additional Quality Assurance work which they should have brought with them, and wait for the sun to shine. In this way, work such as routine auditing of reports, production of other audit reports, or documentary aspects relating to Quality Assurance can be carried out at the site with the utilization of a lap top computer, thus minimizing the time wasted waiting for the ideal conditions to prevail thus allowing this particular audit to be carried out.

One of the key activities is to ensure that the Quality Assurance person is prepared for the audit. This involves understanding the Protocol, knowing the complexities of putting Good Laboratory Practice into the field situation, and knowing where there should be some allowances from the strict laboratory type of GLP that cannot not be enforced in the field. Arrival at the site with the appropriate clothing, bearing in mind the particular hazards that are likely to be met in the field situation that are totally absent in the laboratory, is also another good aspect of the training of that person.

The Report. This is the biggest task, and it requires the transposition of science to paper. Often this must wait until the Autumn or dormant season. Planning, therefore, is the key element.

The Study Director should ensure sufficient times are allowed and be absolutely convinced that all the data are available. This is a natural ongoing process ensuring that all the field data are completed at the time. When it comes to *the final push* to put the data in a meaningful report format, they are all present, duly signed and there are no hidden surprises. For example, when the Protocol had not been followed, the deviations had been identified, written down, and explained at the time of the deviation, and there were no unanswered questions that could arise.

The data itself should be "clean" and checked as the study progresses. The "Editor" wants no surprises, no queries and most importantly no impossible questions to answer about field conditions back at base. In other words: **GOOD FIELD PRACTICE MAKES GOOD REPORTING SENSE, OR GFP = GRS.**

The other solution to this particular problem is to liaise with all of the other parties involved in producing reports to ensure that their time lines meet with yours, and most importantly, that their reports will stand alone and are complete with signatures and other appropriate GLP trappings.

Finally, thought should be given to the reader of the report who is often a regulatory reviewer isolated in an ivory tower, and, therefore, the summary page should not hold any hidden surprises when they delve deep into the body of the report. It should be a well controlled, error free, true story of what happened over that particular experimental period.

The Surrounding Environment. Again, we have identified key problems, the majority of these can be dealt with very early on if adequate planning, communication and liaison are carried out with the surrounding persons and environment.

With regard to crops, rivers and other areas where particular attention should be paid to prevent contamination, this must be discussed at the time, and ideally the site located to ensure minimum disruption to the surrounding environment.

With regard the general public, animals for grazing, etc., this should be again the responsibility of the Study Director to erect where possible a particular restraint, such as fencing, to ensure that if a particular toxic product or an area where minimum disturbance is required then, those beings likely to ramble across the patch should be restricted as far as possible. With regard to the cutting of grass for silage, this must clearly come down to an agreement between the farmer and the Study Director or Principle Investigator to ensure that this is carried out at an appropriate time, and that, where applicable, the material for silage is taken to landfill. Sufficient notice must be given by both parties to ensure minimum disruption to the site and allow the Principal Investigator to remove marker canes, cover up pitfall traps, etc.

Unfortunately, several of the other items identified can not be as easily solved as the little rodent creature who will often find shelter in the weather station control box where it is nice and warm and naturally has an instinct to make a nest thus chewing up immediate material. The weather station, therefore, should be as rodent proof as possible. With regard to other creatures of a biological nature roaming across the site, this basically comes down to prevention and liaison to make sure that trespassing does not happen.

All this sounds good in theory, but in practice it is not always possible. Regular visits to the site are, therefore, essential and a construction of barrier material to be as robust as possible is one of the key elements. However, a tractor spreading biological material or a large, long horned cow will test the most robust system if they are allowed to enter the field site!

Equipment. Unfortunately, all equipment is of a mechanical nature, and, therefore, short of ensuring during the dormant season that these are cleaned, calibrated, serviced and maintained, there is little that the Study Director or Principal Investigator can do to prevent the disasters that we have identified happening in the field situation.

A good supply of spares should always be taken in the field study trailer. The person operating the sprayer, naturally will be well trained but the training also should cover dismantling, trouble shooting, repair and reconstruction. Often, if at all possible, duplicate pieces of equipment should be available at the test site, especially the key materials such as lances, motors/sprayers, nozzles, etc. This situation is totally identical to that in the laboratory environment where, individual pieces of equipment critical to the study must be available for use 24 hours a day.

The GLP requirements of a log book and maintenance record are essential to demonstrate this. Most importantly, the Study Director must see prior to the study, or prior to going to the test site, that the appropriate cleaning, checking, maintenance and calibration have been carried out and are recorded in the laboratory notebook, the field diary, and the equipment log book.

The Regulatory Environment. The final countdown, as has already been explained, is the situation where it is totally outside the control of the ecotoxicology professional. The only solution is to maintain an accurate database on those late evening occasions when you are waiting for the rain to stop. A trip through the Internet to the appropriate web site to review the latest regulatory documents may while away the hours and also enable the Study Director or Principal Investigator to be brought up to date.

In essence, the production of regulatory documents is outside the control of any of the operatives, and, therefore, all possible attempts to keep up to date with these must be made.

Acknowledgments

I would like to take the time to thank my clients for their assistance in the provision of some material and also to thank the reader **for your attention,** in what has been a partly serious and partly light-hearted review of the problems, trials (in both senses) and tribulations of the ecotoxicology professional operating in the current European regulated environment.

I hope that there has been some problem that you have been able to identify yourself with, but most importantly that the possible solutions given may address some of your particular problems.

Literature Cited

1. Council Directive 67/548/EEC of June 27, 1967, Official Journal of the European Communities, No. 196, 16.08.67.
2. Council Directive 91/414/EEC of July 15, 1991, Official Journal of the European Communities, No. L 230, 19.08.91.
3. Council Directive 88/320/EEC of June 9, 1988, Official Journal of the European Communities, No. L 145, 11.06.88.

Chapter 30

Registration of Plant Protection Products: Comparison of the U.S. and EU Models

Héctor F. Galicia[1] and Ronald J. Breteler[2]

[1]Springborn Laboratories (Europe) AG, CH-9326 Horn, Switzerland
[2]Springborn Laboratories Inc., 790 Main Street, Wareham, MA 02571

Registration of plant protection products (PPPs) is going through major changes world wide. The European Union (EU) and the North American Free Trade Association (NAFTA) pursue harmonization of their registration requirements, e.g., efforts in the EU have resulted in the 91/414/EEC Council Directive for the Registration of Agrochemicals. However, success is limited, since only one list of PPPs has been published and its assessment has been delayed by about 4 years. In the Americas, with the passage of the US Food Quality Protection Act (FQPA) in 1996, the concept of a uniform "reasonable certainty of no harm" standard and newly adopted risk assessment criteria will require more than 8,000 tolerances to be reviewed over the next decade. In 1996, the NAFTA technical group for pesticides agreed to pilot joint reviews of applications to register low-risk pesticides. The harmonization procedure must take place at different levels to achieve its objectives. A comparison of the EU and the U.S. models points out some of the factors that ought to be taken into account in the process of mutual acceptability.

Regulatory schemes have been designed to reduce to a minimum the risk that use of man made products represent to individuals. Pressure on government institutions is driven by technological advances and increased public awareness and thus a new perception of risk (*1*). Legislative and regulatory assumptions made years ago are repeatedly challenged as technological advances allow more sensitive measurement of toxic substances to be made. On occasion, novel risks appear, such as, endocrine disrupters and genetically modified food. As the use of science in policy arenas increases, there is a corresponding pressure to enact laws that keep pace with the science upon which the policy is based.

This process has taken place in several countries, and it has resulted in national regulatory law. Globalisation processes and the advantages of the creation of free trade zones, e.g., the EU, NAFTA and MERCOSUR, that have taken place in recent years have promoted the unification of nations having different national regulatory laws. Efforts to harmonize the regulatory procedures are reflected in the 91/414/EEC directive, which aims to regulate the registration and re-registration of new and existing plant protection products (PPPs) in the EU. Similar trends have been observed in North and South America and in Asia. In North America, NAFTA started a program in 1996 for member states to come together to develop requirements and or adopt those already in existence for the regulation of PPPs.

The EU harmonization experience started four years ago and the progress made is a matter of controversy. It is certainly accepted among state members that within the EU a high level of understanding and cooperation have been reached. On the other hand, the objectives set in 1991 have not been fulfilled and still several issues, such as, data protection, remain open to discussion.

The following sections compare the EU and the US regulatory systems and point out at which levels harmonization is possible at the scientific level and at which levels political decisions should take over. Harmonization within the EU and NAFTA ought to be the prerequisite to global harmonization between the two regulatory systems.

Mutual Acceptability

The road to mutual acceptability can be smooth or bumpy and straight or winding. The ease of traverse strongly depends on the similarities and differences of the different regulatory systems being considered. Identification of the likely obstacles in this road will facilitate an analysis as to their potential contribution to mutual acceptability.

The following factors are proposed to be essential in a harmonization procedure and thus should be given high priority, though no claim is made as to the completeness of the selected items. The factors can be of a political and of a scientific nature.

Regulatory Basis: Risk Perception and Legislation. For practical purposes, the discussion is limited to the EU and US regulatory systems; other countries' systems will be discussed in detail in other chapters in this book. Harmonization is certainly a world wide process and thus in Asia, although the Japanese regulatory system is very similar to the EU and US systems, the JMAFF only decided in 1997 to ask for GLP studies for their own reregistration program.

Most of the environmental laws and regulations can be postulated to rest upon a single often non-stated concept, namely the negative default (*2*). This means that for any given environmental risk, it must be assumed that the worst case scenario will happen, and then an attempt has to be made to prove that a less severe situation will actually prevail. In agriculture, the application of a PPP will have effects on the pest(s) to be eliminated or controlled. The non-target crops and organisms are then assumed to be protected from harm. The producer and the formulator of agrochemicals will be asked to prove that at the intended use rate, and given a worst case scenario, the risk involved will be below any unacceptable exposure level.

In order to perform an environmental risk assessment, it has to be accepted that absolute certainty cannot be attained if costs are to be kept at a reasonable level. Present regulatory systems, e.g., EU and US, rely on the premise that it is not possible to give 'proof of zero risk existence', i.e., an absolute zero risk in the use of PPPs. By the same token, the concept of 'zero tolerance' is automatically excluded from present agricultural practices. The latter concept would imply zero use of pesticides in agriculture (no exposure) and thus a zero risk. Integrated Pest and Crop Management practices (IPM and ICM, respectively) are steps being taken in the direction of attaining lower risk. An additional step in the zero tolerance approach is biological control of pests, and the ultimate step is the zero use of PPPs as envisaged by the so called 'biological' (organic) growers. Conseq ently, a regulatory system is self-regulating by the need to make decisions with the highest certainty possible to assure minimal risk. To do this, governments rely on risk assessment, on the one hand, and the cost that an agrochemical company can bear in preparing the data package, on the other. The concept of reducing the use of PPPs so that the risk decreases to acceptable levels has taken the form of mitigation measures, i.e., risk management. A thorough essay on alternative approaches and the extent to which each approach may be useful can be found in (*3*).

Risk assessment consists of a) hazard identification, b) dose-response assessment, and c) the actual environmental risk assessment. Hazard identification is the process of identifying which living being or environmental compartment will negatively be affected by the use of a PPP.

The dose-response assessment is necessary to define the likelihood of an effect to be observed at a higher concentration or an increased exposure. Finally, the environmental risk assessment deals with the associated probability that a substance causes damage and attempts to predict the extent of this probable damage. An additional complication is the relationship between the risk in question and other unknown risk factors, as well as the risk within a selected habitat, e.g., synergy or effect in the population dynamics.

Based on the above, a better approach to "risk assessment" would be a "harm assessment“, which implies having complete and adequate data, and thus an accurate assessment of the harmful effects caused by a PPP could be conducted. Consequently, neither overestimation nor underestimation of the toxic effects could take place. The cost of obtaining (if at all possible) such information is exponentially high, and the time needed to do so probably will be too long for practical purposes. It follows that regulatory agencies are confronted with making decisions from the results of risk assessments with large associated uncertainties. The mistakes that may take place will be made either when defining the hazard or the dose-response relationship. For hazard assessments, false positive and false negative mistakes can be made. The former occurs when a toxic effect is wrongly attributed to a substance, and the latter, when a substance is not attributed a toxic effect although it affects one.

In assessing the dose-response and estimating the need of regulation, the error involved can be defined as overregulation or underregulation of PPPs. Overregulation takes place when a PPP is regulated in accordance with a particular statute too severe for the kind and degree of harm it might cause. The PPP may cause no harm or much less harm than the regulatory authorities may have thought. On the other hand, underregulation will occur when a substance is regulated under a particular regulatory scheme to a lesser degree than it should have been.

A system based on making scientific judgments only when all the sufficient facts are available so no mistakes are made will be inherently slow. Efforts such as these will avoid mainly false positives although false negatives are logically of greater importance.

The US Food Quality Protection Act (1996) (FQPA) (4) adopted a standard with the concept of "reasonable certainty of no harm" and with newly adopted risk assessment criteria. This legislation acknowledges the problem of achieving a 100% certainty, mentioned above, and thus proceeds with a less stringent regulatory system. In this respect, the EU and US systems are following similar objectives, and consequently, they may be considered as comparable at the present time. An effort must be made to develop joint concepts in their assessment models since these models are the fundamental principles in their regulatory thinking. It is essential that their definition of scientific uncertainty and the reasonable certainty of no harm remain similar, otherwise mutual acceptability would become unattainable.

The fact that the EU and US approaches are similar can be confirmed by several recent initiatives between the EU and the US to exchange knowledge and to establish working groups in all disciplines involved in the regulatory process. The experience gained by the US EPA during the reregistration of old pesticides was certainly taken into account by European authorities (*5*). It can be concluded that the present regulatory system and approach are themselves not in question; alternative systems are rarely discussed. In the Food and Drug Administration (FDA) regulatory system, recent proposals for an alternative approach have been issued (*6*).

GLP in the Regulatory Process. Taking into account the previous discussion, an improvement was achieved when it was decided to establish a guarantee for integrity and quality in data packages, and the process is well under way but is not yet complete. The implementation of the GLP principles inherently reduced the uncertainty in the quality of the data submitted, and thus in the risk assessment process, i.e., reducing the likelihood of false positives and false negatives. In principle, an overregulation did not take place because of the introduction of the GLP standards since the risk was reduced without increasing the number or complexity of studies needed but rather increasing the quality of the data collection in the studies themselves.

Both the EU and US regulatory systems, aside from some exceptions, require GLP studies. In the EU not all studies submitted to evaluate a substance will have been performed according to the GLPs; however, for new active ingredients, all studies will have been conducted under GLPs. In the Americas, with the exception of the US and Canada, there are no definitive GLP programs; however, there are initiatives to implement them. The absence of these programs obviously slows down the process of assessment of PPPs in the Americas, and more importantly, it may prevent harmonization and the implementation of uniform principles within NAFTA and between the US and Canada and other nations in the American continent. Alternative roads, i.e., recent efforts to aid international harmonization through ISO guidelines will be addressed in this book. At this stage, it is apparent that the road to mutual acceptability is to pass through the GLP crossroad.

The publication of the revised OECD GLP guidelines (*7*), prepared with an initial input from the US EPA will certainly be a milestone toward mutual acceptance between the US (by extension NAFTA) and the EU. Moreover, it will open the way to the adoption of laboratory accreditation programs in the US, as it has happened in some EU member states and Switzerland. This will ensure better data integrity and higher standards for the agrochemical industry.

A subject of discussion for the future rol⁻ of GLP standards in the regulatory process may be the implementation of these principles in the conduct of risk assessments, as it is now required in the US (*8*).

Test Guidelines. These are essential in the road to mutual acceptability. Countries wishing to take part in an international harmonized system may be better off by not going their own way and investing limited resources in developing their own testing guidelines if the principles of their regulatory systems do not include innovation or allow alternatives to the existing tests. Experience has shown that present regulatory approaches cannot guarantee a virtual zero risk society. These national agencies may profit more if they were to step into the vehicle which is already way far ahead and participate in the efforts taking place in the EU and in the initiatives started in NAFTA.

National regulatory laws have based their requirements on test guidelines, which, in their turn, have been developed at the national level (FIFRA, BBA), in professional societies (ASTM, API, CIPAC), in international organizations (FAO, OECD, ISO, EPPO), or in scientific societies (SETAC, ACS). These organizations have given themselves a mission to generate test guidelines which have been adopted by different regulatory bodies or have been used as a basis for improved guidance.

After analyzing all these test guidelines, it can be stated that there appear to be more similarities than differences among them. Sometimes minor differences in requirements may make impossible the design of a harmonized study protocol. It could be assumed that discussion among the different regulators and organizations, including the agrochemical industry, would produce harmonized guidelines. Unfortunately, this is often not the case, even within the EU.

Freedom to select test guidelines should be preserved for each nation. More important is that studies are not rejected based on the difference in requirements. Regulators, the agrochemical industry, and academia must agree on the interpretations of results from different studies. International expert groups should continue to promote the exchange of information to reach such a level of harmonization. Mutual acceptability should not mean uniformity nor hegemony of one single scientific approach.

Much work has been carried out in the past years on risk assessment, and it would seem that more effort should be invested in the risk assessment schemes presently used rather than in increasing the requirements for more studies for they will marginally aid in improving the regulatory decisions taken.

The Regulatory System in the US. The regulatory system in the US has been described in some detail in previous chapters. A brief overview is presented here.

The Federal Insecticide, Fungicide and Rodenticide ACT (FIFRA). The US law regulating the registration of PPP is the Federal Insecticide, Fungicide and Rodenticide Act (FIFRA) which was enacted in 1947 and amended in 1988 (*9*). FIFRA was originally promulgated to support farmers in order to have efficacious products for agriculture. Now FIFRA's major objectives are the protection of the consumer, the agriculture worker, the producer of the PPP, and the environment.

In 1970, Congress transferred the administration of FIFRA to the newly created Environmental Protection Agency (EPA). Currently, EPA regulates approximately 22,000 pesticide products on the basis of a little over 400 distinct active ingredient groups. In fact, in the US a state may have its own pesticide regulations, as is the case of California with its so called 'state registrations'. The latter consists of a completely separate registration data review system mandated by state law; however, this certification program must still be approved by the EPA, for states cannot have less strict laws than EPA. For example, California's proposition 65, a fairly rigid state initiative dealing with carcinogens, is an example of a very good 'state registration' initiative that was implemented in a fair manner.

In the pesticide regulatory program at the EPA, a rigid regulatory approach has been taken, and it has prevented the EPA from taking a number of innovative approaches to regulation. EPA has labored, chemical by chemical, and submission by submission, over the last two decades, and it still faces major backlogs in reviewing currently registered pesticides. The EU is facing the same problems after 4 years of reregistration efforts.

Residue Tolerances (Maximum Residue Levels, MRLs). Residue tolerances are the basis for export and import of crops and consequently, are a major factor in mutual acceptability. The US residue tolerances, or MRLs in the EU, established by the EPA, are based on the Food, Drug and Cosmetic Act signed in 1938 and amended in 1962 (*10*). The enforcement of the residue tolerances is conducted by the FDA for fruits, vegetables, and grains and by the US department of Agriculture (USDA) for meat, seafood, etc.

Reregistration (FIFRA '88). In 1972, EPA began reregistering PPPs already in the market, using current scientific and regulatory standards. A Registration Standard summarized EPA's evaluation of the available data on an existing chemical, identified and required submission of additional data, and set forth other conditions a registrant had to meet in order for EPA to reregister pesticide products containing the active ingredient. The growing public concern about how long it would take to reassess the potential hazards of existing pesticides, particularly those used on food crops, spurred Congress to enact a series of amendments to FIFRA in 1988 to substantially change EPA's approach to reregistration. Under these amendments, EPA developed a five-phase reregistration process to complete, over approximately a 9-year period, the review of each registered product containing any active ingredient registered before November 1, 1984. This reregistration process has been dubbed "FIFRA-88". This "FIFRA-88" process has obvious similarities to the reregistration process undertaken by the EU in the past five years.

The results of Phase 5 reviews (still in progress) are contained in Reregistration Eligibility Decision (RED) documents. Through the reregistration process, EPA is ensuring that older PPPs meet contemporary health and safety standards, that their labeling is improved, and that their risks are reduced.

The Pesticide Evaluation Process. The EPA is responsible for assessing the risks of all pesticides used or sold in the U.S. to human health or to the environment. The EPA bases registration decisions for new pesticides on its evaluation of test data provided by registration applicants. Applicants for pesticide registration obtain these data from tests that are specified by the EPA. The tests must be performed according to specified protocols.

When a registered pesticide shows evidence of posing a potential human health or environmental safety problem, a Special Review may be triggered. In this Special Review, the pesticide is subjected to an intensive risk/benefit analysis in which all interested parties (environmentalists, manufacturers, users, scientists, the USDA, and the general public) can comment. In a Special Review, EPA may implement various regulatory options to reduce risks associated with a pesticide's use, such as, restricting its use to certified applicators, requiring protective clothing, and prohibiting certain application methods in certain areas only or on certain commodities. EPA may also decide to simply continue its registration if mitigation measures are found unnecessary.

In the mid-nineties, EPA began to recognize the need for harmonized testing guidelines in order to minimize variations among the testing procedures that must be performed to meet data requirements under the Toxic Substances Control Act and the Federal Insecticide, Fungicide and Rodenticide Act. Increase globalization and the initiation of a pesticide reregistration process in the EU lead to the inclusion of guidelines issued by the Organization for Economic Cooperation and Development (OECD) in the guideline harmonization process. To date, only half of the guidelines have been finalized. Twenty five percent more are in draft form. The completion of the harmonized guidelines has been delayed due to enhancement of the Food Quality Protection Act (FQPA).

Food Quality Protection Act (FQPA). Under FQPA, which became effective on August 3, 1996, EPA must also consider the potential for increased susceptibility of infants and children to the toxic effects of pesticides. FQPA requires registered pesticides to meet current safety standards. The Agency also must reassess existing tolerances (MRLs in food), considering aggregate exposure to pesticide residues from many sources and the cumulative effects of pesticides and other compounds with common mechanisms of toxicity. The passage of the Food Quality Protection Act has mandated that a screening and testing program for endocrine disruptors be developed by EPA. Specifically, the FQPA requires that: 1) EPA develop a peer review screening and testing program by mid-1998; 2) EPA implement the program within three years; and, 3) EPA report progress to Congress within 4 years.

The screening and testing program must include estrogenic effects, human health effects, and pesticide active ingredients. However, the screening and testing program could include other reproductive and non-reproductive effects, ecological effects, certain drinking water contaminants, pesticide inert ingredients, anything on the TSCA inventory, and any other environmental agent.

The enactment of the US Food Quality Protection Act in 1996 addresses the above as the concept of a uniform "reasonable certainties of no harm" standard with its newly adopted risk assessment criteria.

In short, the US registration procedure basically consists of the submission of a full data package to the EPA for the active ingredient and the formulated product. Then an iterative procedure is started in which the EPA authorities will ask the applicant to provide them with any additional information on specific concerns, and the applicant's answers are implemented into the evaluation processes. Upon complete review of the data, which may take up to about 2 years, the final decision is made on product registration with its intended uses and the residue tolerances. The resulting decision is published in the Federal Register, which is the US official journal.

The US registration procedure is thus an iterative process which normally involves queries from the EPA on the full data package (issue of dossier completeness in the EU). Responses are expected from the scientists who conducted the studies or the representatives of the agrochemical producer (Figure 1).

The Regulatory System in the EU. The regulatory system in the EU has been presented in detail in other chapters in this book. Therefore, in this chapter, only the highlights will be pointed out.

The Registration Procedure and the 91/414/EEC Commission Directive. Commission Directive 91/414/EEC (*11*) set the path to the reregistration effort in Europe. This commission directive includes Annexes II A and II B for chemical and microbial PPPs, respectively; as well as Annexes III A and III B for their corresponding formulated products. The main step in the registration of a PPP, according to this directive, is the periodic publication of a list of active ingredients which are to be reassessed. The agrochemical producers communicate to the rapporteur country that they will submit a complete dossier containing the data requirements for the inclusion of the active substances into Annex I.

Figure 1. Comparison of the US and EU Registration Models.

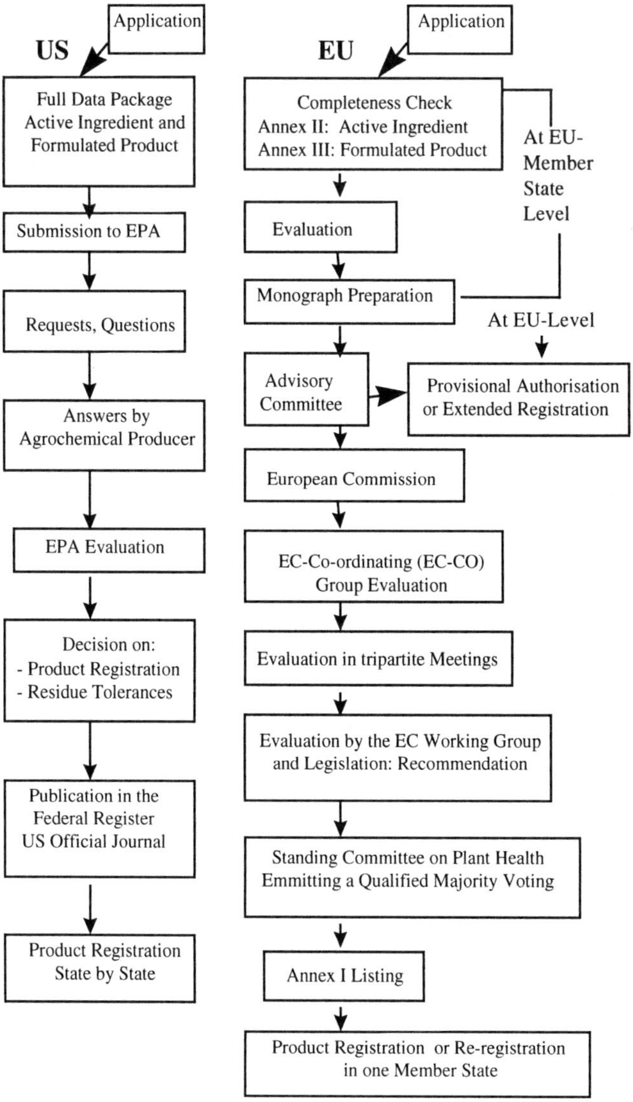

Thereafter, the regulatory authorities of the corresponding rapporteur country conduct a completeness check and if satisfied, they proceed to an evaluation of the data submitted. Once all data requirements have been answered, including additional queries or demand for higher tier studies, the rapporteur country prepares and submits a monograph for that active ingredient to the advisory committee. At this stage, a request for provisional authorization may be granted. The advisory committee takes the case to the European Commission where an ECCO (EC-Coordinating) group evaluates the findings, and through tripartite meetings and also at the EC Working Group level, the evaluation is terminated. A recommendation is given to the Standing Committee on Plant Health, which reports a qualified majority vote. If the decision is positive, the active ingredient is listed in Annex I, and product registration or reregistration for sale in one member state can proceed (Figure 1). This procedure, in contrast with FIFRA '88, strongly depends on the conclusions and risk assessments conducted by the applicant.

Other Stops on the Road to Mutual Acceptability. Discussion of the experiences gained by the EPA and the EU in the past few years has pointed out the need to concentrate on several that need attention. Some of these issues are noted in the following paragraphs:

The National Regulatory Structures. Some of the member states in the EU and most of the countries in Latin America will struggle to implement similar regulatory systems as those established by the US, UK and Germany (*12*). Their staffs are either limited in number or have not been exposed to similar evaluation schemes. Implementation of GLP accreditation programs, evaluation of test guidelines and assessment of data packages, all at the same time, must involve very careful planning and use of resources. The EU experience has shown that this may lead to a concentration of power in a few, better prepared member states. The contrasts among countries in the Americas are more pronounced, and the danger of absorbing a nation's regulatory process by an established, stronger one, is imminent.

Applicants and Data Protection. The EU experience has shown that a difference exists between the approach followed by multinational and local agrochemical companies, or by proprietary and generic producers. A likely harmonization between the EU and the US will have to take into consideration data protection, since mechanisms are different between the two regions.

Data Interpretation - Comparison of Zones, Regions and Scenarios. The larger the trade zone where harmonization is meant to take place the more different the geographic and climatic regions and, therefore, the more scenarios needed to be considered. Mutual acceptability in the EU, based on the proliferation of climatic scenarios resembling more those found in Northern Europe, will pose problems to state members located in the southern part of the EU.

The extension can be made to the use of soil types, sediments, surface waters, groundwater, aquatic and terrestrial organisms, birds (yearly migration), and crops (e.g., bananas in Spain, Canary Islands and Mexico). It is certainly difficult for an applicant to perform a study with a soil that may fully correspond to his application but that may not match the standards set by the regulatory bodies. Within the EU, the mutual recognition of authorizations is in the implementation process and will address the following sectors: Efficacy, Human Health (exposure to the PPP), Human Health (exposure to PPP residues), Fate and Behavior in the Environment and Impact on Non-target Species (*13*). It is expected that the differences in all species and abiotic components of an ecosystem will be overcome by choosing realistic worst cases and not those of abuse. The process requires preparation of working documents with review and revision. This aspect is certainly one of the most difficult ones since it presupposes trust in the PPP assessment by other member state. Mutual recognition between the EU and the US would certainly profit from the experience to be gained in the EU.

EU and US Regulatory Systems: Differences and Similarities. Some common features and differences are:

- The objective of the EU and US systems is to reduce false positive and false negative mistakes, to increase the certainty in the risk assessment, and to avoid under- or overregulation.

- The present development may give the impression that the EPA may have started to follow a more pragmatic approach: "reasonable certainty of no harm", whereas the EU may be following the strategy of acquiring more data to reduce uncertainties. The question, in fact, is which regulatory system at present demands more studies, and how are they evaluated?

- In the EU and in the U.S., the registration process is centrally organized (Brussels, Washington). However, in the EU it is not centrally evaluated. In the EU, each member state prepares a monograph for an allocated substance. In the US, a similar situation takes place in the states which have their own registration data review, as is the case of California.

- In the EU, each member state will accept the requirements set by the EU unless the directive 91/414/EC has not been implemented into their law, as is the case in Germany. In the US, a state may have its own 'state registrations' (e.g., California) and thus may not necessarily accept an EPA decision. California has a separate registration data review system that is mandated by state law; however, it still needs to be approved by the EPA.

- Data protection in the US provides for a 15-year data compensation for all data submitted to the EPA, whereas in the EU,there are 5 years' data protection for data concerning the active ingredient and no additional protection for supplementary data submitted for the formulated product.

- In the EU, industry is asked to submit a dossier including risk assessment and recommendations on the substance of interest; whereas in the US, this risk assessment is performed by the EPA authorities themselves (obviously agrochemical producers in the US conduct their own risk assessment anyhow).

- Both the EU and the US have as their top priority sound scientific judgment for evaluation of the data for each PPP and follow the so called scientific approach.

- One of the most difficult areas for evaluation is the area of carcinogenicity, and this will be one of the essential factors in mutual acceptability and, consequently, a main topic in the harmonization efforst between the EU and the US.

Conclusions

The EU and US systems are considered to follow similar objectives, and they may be considered to be similar at the present time. It was pointed out that an effort must be made to develop joint concepts in the assessment models in order to have common, fundamental principles in their regulatory thinking. It is essential that their definition of scientific uncertainty and the reasonable certainty of no harm remain similar, otherwise mutual acceptability would become unattainable.

The implementation of GLP principles into regulatory testing may not have reduced the scientific uncertainty in the data submitted; but, in combination with study guidelines, it has eliminated unreliable data and thus aided the risk assessment process for the registration of PPPs. Therefore, the inclusion of GLP requirements in harmonization procedures can only be desirable.

A major aspect for mutual acceptability is that studies should not be rejected based on the difference in national testing requirements. Regulators, the agrochemical industry, and academia must agree on interpretations of results from the same type of study with slightly different testing requirements. It is suggested that international expert groups continue to promote scientific exchange to reach such a level of harmonization.

It has been assumed here that the input of the US and Canadian regulatory systems will be the driving force in the NAFTA effort, and thus by extension, the argument presented for the US EPA ought to be valid for the North American continent. Other regulatory authorities on the continent will profit from the disclosure of these efforts. A global harmonization, EU-NAFTA, is now an ambitious vision that many contemplate as the ultimate step. If the harmonization processes within the EU and NAFTA are as smooth as could be expected, global harmonization will be attainable before long.

Acknowledgments

We wish to thank the reviewers for the useful comments to improve the contents presented herein.

Literature Cited

1. Townsend, Joseph B. *Responding to Regulatory Changes in Agrochemical Research; in Garner et al. (Eds.) "Good Laboratory Practice Standards: Applications for Field and Laboratory Studies",* ACS Professional Reference Book; American Chemical Society: Washington, DC, 1992; Chapter 1, pp 3-10.
2. Wilson, Albert R., 1991. *Environmental Risk: Identification and Management.* Lewis Publishers: Chelsea, Michigan, 1991, Chapter 1, pp 3-27.
3. Cranor, Carl L. *Regulating Toxic Substances;* Oxford University Press, Inc.: New York, NY, 1993, 252 pages.
4. *"1996 Food Quality Protection Act: Implementation Plan"*, United States Environmental Protection Agency; Prevention, Pesticides and Toxic Substances; March, 1997; Part 4.
5. Gardiner, J. A., 1990. *The USA EPA Experience; in Thomas B. (Ed.) "Future Changes in Pesticide Registration Within the EC";* BCPC Monograph No. 44; pp. 63-67.
6. Krauss, Michael I., 1996. *"Loosening the FDA's Drug Certification Monopoly", the Competitive Enterprise Institute (CEI's Publication): Breaking the FDA's Drug Approval Monopoly, Implications for Tort Law and Consumer Welfare,* Health Care Reform Project; ISSN#1085-9055, 25 pages.
7. "OECD Series and Principles of Good Laboratory Practices and Compliance Monitoring". OECD, Principles on Good Laboratory Practices (as revised in 1997). Environment Directory, OECD, Paris 1998.
8. Garner, Willa. Personal communication, 1998.
9. "The Federal Insecticide, Fungicide, and Rodenticide Act as Amended", U.S. Environmental Protection Agency, Office of Pesticide Programs, U.S. Government Printing Office: Washington, DC, 1989, 617-003/84346.
10. Food, Drug and Cosmetic Act, 1906, 1938 amended 1958, 1960, 1962, 1968. 21 U.S.C.
11. Council Directive 91/414/EEC of 15 July 1991, Official Journal of the European Communities, No. L 230, 19.08.91.
12. Thomas, Barry. *"EU Registration: Have We Reached the Cross-roads?",* presented at the IBC Conference on 'Registration of Agrochemicals in Europe', 1997; 8th & 9th October, London, UK.
13. Lynch, Mark R., 1997. *"Mutual Recognition of Authorizations",* presented at the IBC Conference on 'Registration of Agrochemicals in Europe', 8th & 9th October, London, UK.

Chapter 31

Good Laboratory Practices and Pesticide Regulation in Japan

Fumihiko Ichinohe

Agricultural Chemical Inspection Station, Ministry of Agriculture, Forestry and Fisheries, Tokyo 187, Japan

In Japan, the Agricultural Chemicals Regulation Law provides that "to obtain registration, applicants shall submit the test results concerning effectiveness, phytotoxicity, toxicity and persistence in crops and soil, etc., of each agricultural chemical." Most test guidelines for these studies follow the OECD Test Guidelines, and the rest of them are now considered to be harmonized with OECD guidelines. For toxicity studies, we notified applicants about the "Good Laboratory Practice Standards for Toxicological Studies on Agricultural Chemicals" and required them to comply with GLP standards for all study data generated. In accordance with the OECD Council Recommendation, we will gradually extend the scope of types of study data for the application of the GLP system after the establishment of domestic test guidelines. The GLP standards of Japan are almost similar to OECD GLP Principles and the GLPs of the US. Japan accepts the inspection results of the countries with whom Japan has concluded a bilateral arrangement.

GLP System for Agricultural Chemicals in Japan

The GLP system is a very effective means to secure the reliability of study data. In June 1979, the US Food and Drug Administration (*1*) took the initiative in introducing this system for pharmaceutical products. In May 1981, the Organization for Economic Cooperation and Development (*2*) established principles to help individual countries bring their GLP standards up to an international level. Based on international trends, the Ministry of Agriculture, Forestry and Fisheries also had started considering the introduction of this system. On August 10, 1984, the Director-General of Agricultural Production Bureau issued the notice entitled "the GLP Standards for Toxicological Studies on Agricultural Chemicals" (*3*) which took effect on October 1, 1984. Since then, the GLP system is applied only to toxicological study data. In establishing a GLP system for agricultural chemicals, the Ministry thoroughly considered other GLP systems and related information in Japan and abroad, such as actual situations at domestic testing laboratories, the GLP system for pharmaceutical products introduced by the Ministry of Health and Welfare, and the US GLP system and how those systems work. The Japanese GLP system is well harmonized with foreign countries.

Bilateral Agreement Concerning GLP Systems. As many countries had introduced GLP systems, reliability of toxicological study data prepared in a foreign country could be secured, and those data could then be used for the registration application of agricultural chemicals. However, it is very difficult to verify the reliability of study data generated by laboratories in foreign countries with the way studies are conducted in Japan. Consequently, it is efficient to make bilateral agreements between the two countries concerned saying that each government confirms and assures that the laboratories involved comply with the GLP standards applied in the country. So far, Japan has made bilateral agreements with the four countries listed in Table I, and toxicological study data are now mutually accepted in those countries.

Table I. Japan's Counterparts in Bilateral Agreements

	Date of Agreement
U S	Sept.16, 1987
U K	Oct. 7, 1987
Germany	Feb. 16,1988
Switzerland	Jan. 18, 1993

Japan also accepts toxicological study data prepared by laboratories in countries with whom Japan has not yet made an agreement. In those cases, the GLP standards of the country are required to comply with OECD GLP Principles (*2*), and GLP compliance by the laboratory for study data it generates should be certified by the appropriate authorities in the country.

Mechanism of the GLP System. The purpose of the GLP system for agricultural chemicals is basically to confirm the reliability of individual data. In practice, a system of "the confirmation of GLP compliance" includes a method for inspecting the laboratory organization and its operation, as well as other related matters. Under this system, laboratories are required to apply every three years for an inspection under which the laboratory is checked to see if it has been operated in compliance with the GLP standards for the last three years. Laboratories preparing toxicological study data are required to undergo this inspection for the registration application of agricultural chemicals.

Application for a Reliability Confirmation Inspection. Table II shows the number of applications submitted and inspections conducted for last 13 years, from 1984 (when the GLP system was introduced into Japan) until 1996. Roughly, some 40 domestic laboratories and 50 foreign laboratories have submitted inspection applications in each three-year period. Most of the foreign applications came from the four countries that have obtained bilateral agreements with Japan. Others came from laboratories in France, the Netherlands, and Italy.

Table II. Applications and Inspections for Confirmation of GLP Compliance

	Japan		*Overseas*	
Year	*Applications*	*Inspections*	*Applications*	*Inspections*
1984	7	1	1	0
1985	21	11	9	0
1986	7	10	17	0
1997	6	10	12	0
1988	14	15	15	0
1989	12	8	20	0
1990	11	12	17	0
1991	19	18	14	0
1992	12	13	27	0
1993	11	9	11	0
1994	18	14	13	1
1995	14	11	27	0
1996	11	7	10	0

With the application, on-site inspections are carried out at domestic laboratories that have conducted toxicological studies for registration applications of agricultural chemicals within the last three years. During the interval of three years, some 35-40 laboratories undergo this inspection. For foreign laboratories, an inspection was conducted in 1994 in a U.S. laboratory in collaboration with the U.S. Environmental Protection Agency under the bilateral agreement.

Future of the GLP System. The GLP system can be applied not only to toxicological studies but to other various studies. Some OECD countries have already applied the GLP system to studies other than toxicological studies, such as physical/chemical property studies. The scope of the studies under the GLP system will surely be expanded in Japan.

Under such a global tide, the Ministry of Agriculture, Forestry and Fisheries (MAFF) now takes into consideration the expanding of the types of studies to which the GLP system will be applied in response to requests from foreign countries. MAFF will apply the GLP system to physical/chemical property studies and ecotoxicological studies on aquatic organisms before long. In the future, this system will be applied to other various studies.

Pesticide Safety Evaluation in Japan

Pesticides are agricultural chemicals that are used for the control of insects and fungi which injure agricultural crops and/or for the improvement and control of physiological functions in the plant. The ideal is that pesticides will be effective only against pests and will be harmless to people, useful animals and plants, and the environment. However, they have physiological properties that may show injurious effects to the latter. Therefore, when we deal with pesticides, we must pay appropriate attention to human health and environmental conservation.

To this end, it is necessary to know exactly the properties of the pesticide in question in order to make a suitable safety evaluation.

Safety Evaluation System. Of all chemicals, pesticides, as a group, have the most extensive safety and toxicological data bases. Safety evaluations are required for all pesticides prior to registration and marketing. These tests are designed to mimic the potential routes and conditions of exposure for man and the environment. According to the Agricultural Chemicals Regulation Law (*4*), no manufacturer or importer is allowed to provide a pesticide for sale on the domestic market in Japan unless registration has been granted by the Minister of Agriculture, Forestry and Fisheries. In the process of registration review, safety evaluation is strictly implemented.

Procedures for Registration Application. For the registration application, the manufacturer or importer of agricultural chemicals makes an application by submitting a product registration application form, experimental data, and a sample of the product. These materials are submitted to the Minister of Agriculture, Forestry and Fisheries via the Director of the Agricultural Chemicals Inspection Station.

Product Registration Application Form. This application form is based on supporting data that are described as specified below:

- Name and address of the applicant for registration
- Common name of the pesticide and its trade name
- Physical and chemical properties
- Name and content of active ingredients
- Name and content of other constituent
- Kind and material of containers or packs and net content
- Spectrum of pests (including weeds and rats) controlled by the products and the manner of application. For plant growth regulators, spectrum of crop and purpose of application.
- Mammalian toxicity and detoxication method
- Toxicity to aquatic organisms
- Flammability, explosiveness and skin irritability

- Precautions necessary for storage and/or use
- Name and location of manufacturing plant
- Manufacturing process and names of the people in charge of manufacturing

Data Requirements. Listed below are the required studies for pesticide registration:

- Efficacy and phytotoxicity study – A phytotoxicity study is not only for target crops, but also for adjacent and/or subsequently raised crops.
- Toxicity study – Toxicological data requirements are described in Table III. The toxicity test must be conducted properly and impartially, otherwise safety evaluations are not secured with accuracy and reliability. These data should be produced in laboratories that are confirmed by the authorities as being in compliance with the GLP standards.
- Residue chemistry --

    1) Residue remaining in the crop – Information on the amount, frequency and time of pesticide application and the results of tests on the level of residues persisting in the treated crop are required for pesticides for food use. Trials are conducted at more than two different Prefectures and samples are analyzed at more than two separate laboratories. A description of the analytical method also is required.

    2) Retention in the soil – Information on the retention levels of the pesticides in soil is required. This study is conducted in the field and in pots.

- Physical and chemical properties of the active ingredient(s)
- Composition of the technical grade of the active ingredient(s) – Impurities have to be identified and quantified with the most advanced analytical techniques.
- Formulation – The kinds and contents of any inert ingredient in the formulation have to be made clear.
- Environmental impact on fish or other nontarget organisms including silkworms, bees, natural enemies and wild birds.

Review of the Data for Registration. Review is conducted by checking whether the results of the studies submitted in the application fall under the conditions laid down for withholding a registration. If they fall under even one of the registration withholding clauses, the applicant is required to amend the statement entered in the application form and/or to improve the product quality, otherwise the registration is withheld.

The conditions for withholding registration are as follows:

- False facts are found in application statement.
- Crops are damaged by the product that is used according to the directions given in the statement.

Table III. Toxicology Data Requirements for the Registration Application of Pesticides

Kind of Data Required	*General Food Crop*	*Use Pattern Non-food Crop*	*Test TGAI* [a]	*Substance End-use Product*
(Acute toxicity study)				
Acute oral toxicity	R	R	R	R
Acute dermal toxicity	R	R	R	R
Acute inhalation toxicity	R	R	R	R
Primary eye irritation	R	R	N	R
Primary dermal irritation	R	R	N	R
Dermal sensitization	R	R	N	R
Acute delayed neurotoxicity	R	R	R	N
(Subchronic toxicity study)				
Subchronic oral toxicity	R	R	R	N
Subchronic dermal toxicity	C	C	C	N
Subchronic inhalation toxicity	C	C	C	N
Subchronic neurotoxicity	C	C	C	N
(Long term toxicity study)				
Chronic toxicity	R	C	R	N
Oncogenicity	R	C	R	N
(Special toxicity study)				
Reproduction	R	C	R	N
Teratogenicity	R	R	R	N
Mutagenicity	R	R	–	–
(Others)				
Metabolism	R	R	–	–
Pharmacology	R	R	–	–

[a]Technical grade of active ingredient

R: Required

C: Conditionally required

N: Not required

– : Not specified

- The product can do harm to man and livestock even if prescribed means are taken to avert such damage.
- *In view of the persistence of residues in crops, the product contaminates crops and can do harm to man and livestock.
- *The product contaminates the soil and can do harm to man and livestock due to ingesting any crops that are raised in the contaminated soil.
- *In view of its toxicity to aquatic animals and plants and its toxicity persistence, the product pollutes water to cause damage to them and can do serious harm.
- *The product pollutes water in the water catchment area for public use and can do harm to man and livestock due to use of that water.
- The trade name of the product can cause misunderstanding about the main components and/or their effectiveness.
- The product is so inferior in efficacy that it should not be approved as an agricultural chemical.
- For the product to which the official standard has been applied, it does not fit in with official standards and its efficacy is inferior to other products conforming to the standard.

The criteria of the conditions from the asterisked items, above, are laid down and noticed by the Director-General of the Environmental Agency. The criteria make up the so-called "registration withholding limit", which is a kind of tolerance. In particular, concerning residues remaining in a food crop, all data on toxicity and residual persistence are strictly reviewed by the authorities concerned, and these limits for each pesticide ingredient are noticed for the food group under which crops for the registration application are assigned. At the Agricultural Chemicals Inspection Station, the registration application is under review to ascertain whether the residual level remaining in the crops exceeds the registration withholding limit when the product is applied with regard to the directions for use stated in the application form.

Regarding the third condition, above, for safety in use, acute toxicities mainly are evaluated by the authorities concerned, and those evaluations are reflected in the registration review. Those registration applications, which passed this strict review, are set with proper application directions such as time, frequency, rate and protective clothing and then registered in the name of the Minister of Agriculture, Forestry and Fisheries. These items are indicated on the label of the container.

Designation of Toxic Substances. Pesticide products, for which their acute toxicity is very important, are under review in the Ministry of Health and Welfare and are designated as poisonous or deleterious substances under the Poisonous and Deleterious Substances Control Law (*5*). Designated pesticides are then indicated to that effect in a statement on the label and are requested to be kept in a locked place.

With regard to the safety of pesticides, we also have to examine other properties of these chemicals. Their properties include phytotoxicity; pollution to the environment, such as, rivers, lakes and sea; and injury to people and useful animals. With a view to averting such damage, many kinds of data must be submitted for the registration application and these are under strict review. Consequently, registrations are only granted under certain terms and conditions.

Establishment of the Tolerance. Tolerances of pesticides constitute two groups. One group is composed of tolerances based on the Food Sanitation Law (*6*), which is called "the pesticide tolerance". The others are tolerances determined by the Director-General of the Environment Agency (*7*) to be based on the Agricultural Chemicals Regulation Law (*4*), which is called "the pesticide registration withholding limit". The latter tolerances are set up based on those that the former has not determined.

The principle for setting up a tolerance is as follows: a tolerance is laid down within a range so that the total dietary exposure level does not exceed the ADI. The following formula gives the relationship between the tolerance and the ADI:

$$\Sigma \left(\begin{array}{l} \text{Average daily intake of} \\ \text{each food group (kg)} \\ \times \text{ tolerance (ppm)} \end{array} \right) = \begin{array}{l} \text{Estimated dietary} \\ \text{exposure level} \\ \text{(EDEL) mg} \end{array}$$

$$\text{EDEL} < \text{ADI} \times 50 \text{ mg/man}$$

The tolerance is, in principle, set up for each food group. If it is necessary to set up a tolerance for many food groups, the total dietary exposure is estimated from the sum of the intake by individual food groups, which should not exceed the ADI × 50 mg/man. The amount of intake by food groups is derived from the Current Report of National Nutrients of the Ministry of Health and Welfare.

Pesticide residue data are also taken into consideration. The tolerance is set up with a certain safety margin calculated from the residue level obtained from the data. The relationship between the tolerance and residue level is given in the following schematic:

$$\text{Tolerance} > \text{Maximum residue level} > \text{Average residue level}$$

On setting up tolerances, the estimated total dietary exposure is supposed to be based on the following assumptions and conditions:

- The pesticide is applied to all varieties of crops of each food group for which the tolerance is being set.
- The residue level remaining in the crops is equal to the tolerance.
- Any decrease of the residue level by washing, cooking and processing is not taken into consideration.

From these assumptions, the total dietary exposure level is estimated as being much more than the real intake level. The reasons are as follows:

To the first assumption, the varieties of the crop, for which registration is granted, do not all come under the food group, but only some part of it. Furthermore, the pesticide concerned is not necessarily the only one which is used on the crops, as some other pesticides might be applied to the same crops as well.

To the second assumption, the average residue level is much less than the tolerance.

To the third assumption, crops are, in general, washed, cooked and processed, thus removing much of the pesticide residues.

Toward the end of promoting global trade of agricultural commodities, it is necessary to set up food and feed standards. FAO/WHO of the United Nations has jointly established the Codex Alimentarium Committee (CAC). Pesticide residues of food and livestock feed among the countries are covered by the Codex Committee of the Pesticide Residues (CCPR) who establish the Maximum Residue Limits (MRL) with the support of the Joint Meeting of Pesticide Residues (JMPR).

Japanese tolerances are set up not to exceed the ADI so people do not get damaged at all, even though a person may get a daily intake of the pesticide residue over his lifespan. Meanwhile, MRLs of the CCPR are set up based on the Good Agricultural Practices (GAPs). MRLs are attributed to the level of the pesticide residues remaining in the agricultural products raised under the best control of cultivation with the application of the minimum amount of pesticide. Accordingly, MRLs are established apart from the ADI. In Japan, tolerances are directly established in view of human health, but MRLs are primarily set in view of the necessity of the pesticide for agricultural production. These MRLs seem to be separate from the levels of pesticide residue in food and feed that are related to human health. However, CCPR has conducted a total diet survey and/or market basket survey using residue of our agricultural chemicals and has confirmed the estimated total level of pesticide residue intake to be far below the ADI.

Setting Up the Directions for Use. Residue levels are subject not only to weather conditions but also to many other factors, such as, the application time, frequency of application, rate, and methods, such as, top dressing or mixing in the soil. Accordingly, on the basis of data of pesticide residue levels in crops, the directions for use are set up in such a way that the residue levels do not exceed the tolerance. The directions include the applicable varieties of the crops, rate, time (the minimum number of days from the last application before harvest), frequency, and precautions for application, etc. These items are indicated on the label of the containers.

Table IV shows some examples of setting up tolerances and directions for use of a registered pesticide that is applied to paddy rice and several crops from the group of fruits and fruiting vegetables. In this instance, the estimated total dietary exposure level is far less than the ADI $\times$ 50.

Aquatic Assessment – Risk Assessment/Risk Mitigation. Aquatic risk assessment is conducted in Japan from two aspects; 1) risk to human health via drinking water, and 2) risk to aquatic organisms.

Table IV. Example for Setting Up the Tolerances and Directions for Use

Food Group	Food Factor (g)	Tolerance (ppm) (MRLs)	Estimated Dietary Exposure Level-mg (TMDI)	ADI x 50 (mg/man)	Maximum Residue Level (ppm)	Direction for Use
Rice	203	5	1.015		1.7	2 times, 14 days before harvest
Large Fruits Class2	68	5	0.340		2.1	Apple, 3 times, 7 days before harvest
Fruiting Vegetables Class2	42	10	0.420		3.2	Cucumber, 2 times, 1 day before harvest
Total			**1.775**	**10**		

Assessment and Mitigation of Risk to Human Health via Drinking Water. The main exposure route is run-off from paddy water. Spray drift is taken into account qualitatively, but the calculation of the Predicted Environmental Concentration (PEC) is not done.

To assess risk to human health caused by a pesticide via run-off, it is necessary for applicants to submit a dissipation study of the pesticide in water from two small scale paddy fields with different soil types. Normally, that is the 7 to 14 day dissipation study in which 7 days' average concentration level of the pesticide is obtained. The 150 days' average concentration level as PEC is estimated from seven days' average concentration level. The reason for taking a 150 days' average concentration level is that the usual time period for filling a paddy field is about 150 days in Japan. The PEC is evaluated in order to compare it with the regulatory withholding standard concerning water pollution (cut-off criteria level). The cut-off criteria level is established on the basis of the ADI of the active ingredient concerned. If the PEC exceeds the cut-off criteria level, the registration of the pesticide is withheld and a risk mitigation measure, such as a reduction in the application rate or an establishment of a flow-out stopping time of paddy water has to be considered.

As mentioned above, for spray drift, PEC is not calculated in Japan, however, this exposure route is qualitatively taken into account and, when necessary, a risk mitigation measure is taken by the description of the precaution on the label.

For risk mitigation at the post-registration stage, the following measures are taken when necessary: The Standards for Safe Use of Agricultural Chemicals Concerning Prevention of Water Pollution (*8*) provides that users of the pesticides, for which the Environmental Quality Standards for Water Pollution (*9*) are established, shall pay attention not to allow pesticide spray to drift over rivers and where water purification plants are located. The Standards for Safe Use of Agricultural Chemicals Concerning Aerial Application (*10*) provides that the aerial application shall not be carried out in areas where rivers and water purification plants are located.

A pesticide that may cause water pollution in a public water area and, consequently, may cause damage to man and livestock when used in large quantities in extensive areas, may be designated by the Government as an agricultural chemical that causes water pollution. For a pesticide so designated, the Prefectural authorities may determine the area, when necessary, where any user of the pesticide shall not use the pesticide without obtaining a permit in advance from the Prefectural authorities.

Assessment and Mitigation of Risk to Aquatic Organisms. The LC_{50} of common carp (48hrs) and LC_{50} of *Daphnia* (3hrs) are used for the assessment. The cut-off criteria level is established on the basis of the LC_{50} of common carp (48hrs). This cut-off criteria does not consider the PEC. However, for pesticides that do not exceed the cut-off criteria, a risk assessment is conducted that takes into consideration the PEC. In this case, the PEC is calculated from direct overspray to a body of water with a depth of 5 cm. According to the result of the risk assessment, a risk mitigation measure is taken.

For spray drift, the PEC is not calculated in Japan; however, this exposure route is qualitatively taken into account and, when necessary, a risk mitigation measure is taken (e.g., the precaution on the label).

As one measure of risk mitigation, precautions for the users are described on the label. An example of a precaution is “This pesticide may cause damage to fish and crustaceans and may not be used near nurseries of fish and crustaceans”.

For risk mitigation at the post-registration stage, the following measures can be conducted when necessary: the Standards for Safe Use of Agricultural Chemicals Concerning Prevention of Damage to Aquatic Animals (*11*) provides that users of the pesticides designated by this Direction shall pay attention to assure no drift of the applied pesticide over rivers, lakes, ponds, etc.

A pesticide that possesses the potential for serious damage to aquatic animals and plants when used in large quantities in extensive areas may be designated as an agricultural chemical that causes water pollution by the government. For such designated pesticides, the Prefectural authorities may determine the area, when necessary, where any user of the pesticide shall not use the pesticide without obtaining a permit in advance from the Prefectural authorities.

Literature Cited

1. *Good Laboratory Practice for Nonclinical Laboratory Studies;* Food and Drug Administration; Code of Federal Regulation No. **21**, Part **58**; U.S.A: **1979**.

2. *OECD Principles of Good Laboratory Practices;* C (**81**) **30** (Final) Annex; OECD: Paris, **1981**.

3. *Good Laboratory Practice Standards for Toxicological Studies on Agricultural Chemicals;* Director-General of Agricultural Production Bureau, Ministry of Agriculture Forestry and Fisheries; **59** Nosan No. **3850**; Japan: **1984**. (in Japanese).
4. *Agricultural Chemicals Regulation Law;* Ministry of Agriculture, Forestry and Fisheries; Law No. **82**; Japan: 1984. (in Japanese).
5. *Poisonous and Deleterious Substances Control Law;* Ministry of Health and Welfare; Law No. **303**; Japan: **1950**. (in Japanese).
6. *Food Sanitation Law;* Ministry of Health and Welfare; Law No. **233**; Japan: **1947**. (in Japanese).
7. *Standards for Withholding Agricultural Chemicals Registration;* Director-General of Environmental Agency; Ministerial Notification No. **37**; Japan: **1978**. (in Japanese).
8. *Standards for Safe Use of Agricultural Chemicals Concerning Prevention of Water Pollution;* Minister of Ministry of Agriculture, Forestry and Fisheries; Announcement **6** Nosan No. **1623**; Japan: **1994**. (in Japanese).
9. *Environmental Quality Standards Regarding Water Pollution for Protection of Human Health;* Director-General of Environmental Agency; Ministerial Notification No. **16**; Japan: **1993**. (in Japanese).
10. *Standards for Safe Use of Agricultural Chemicals Concerning Aerial Application;* Minister of Ministry of Agriculture, Forestry and Fisheries; Announcement **6** Nosan No. **1623**; Japan: **1994**. (in Japanese).
11. *Standards for Safe Use of Agricultural Chemicals Concerning Prevention of Damage to Aquatic Animals;* Minister of Ministry of Agriculture, Forestry and Fisheries; Announcement **6** Nosan No. **1623**; Japan: **1994**. (in Japanese).

Chapter 32

GLP National Status and Facilities in India for Pesticide Product Registration

B. Vasantharaj David

Jai Research Foundation, P.O. Valvada—396 108 Gujarat, India

This paper defines the principles of Good Laboratory Practice (GLP) and traces the history of GLP worldwide. In India, the usage of synthetic pesticides commenced in 1948. The history of the enactment of the Insecticides Act in 1968 and the Rules in 1971 has been outlined. In 1978, the subcommittee on pesticide toxicology was constituted under the chairmanship of Dr. B. B. Gaitonde, and it finalized the guidelines for data requirements for registration of pesticides in India.

The concept of GLP in India was initiated in 1983. The necessity for GLP laboratories in India was felt to promote export for Indian companies and also make available data generation to GLP at a reasonable price. GLP facilities in India established in 1992 by Rallis Research Centre and in 1996 by Jai Research Foundation have also attracted the global market. The problems and challenges in implementing GLPs in India and the prospects of GLP facilities in India are discussed.

History

Good Laboratory Practice (GLP) basically aims at the philosophy of laboratories carefully documenting all activities of designing, performing and reporting safety studies on chemicals and preparations under strict monitoring by an internal independent quality assurance system thus enabling reconstruction of the studies at any time afterwards (*1-2*).

Though laboratories observed 'Good Laboratory Practice' for years, only in the early 1970s its application to control laboratories by the Government began. In 1973, New Zealand was the first country to promulgate the Testing Laboratory Registration Act, and in March 1973 similar legislation came into force in Denmark.

In August 1976, the FDA released a draft set of GLPs and the proposed GLP regulations were published on 19 November 1976 in the Federal Register. The final GLP regulations were published in the Federal Register on December 22, 1978, which became a legal entity in the U.S. on 20 June 1979. On 24 October 1984 changes to the US GLPs were proposed and published as revised good laboratory practice regulation on 4 September 1987 in the Federal Register entitled "Good Laboratory Practice Regulations, The final Rule".

However, proposed GLP standards relating to pesticides were published by EPA under FIFRA in April 1980 (45 FR 26373). EPA Final Regulations were enforced in November 1983 (FIFRA : 40 FR 53946). In 1987 the proposed Amended Regulations under FIFRA were issued as 52 FR 48920 (*9*).

In 1978, the OECD set up a special program on the control of chemicals. One of the first priorities of this program was the creation of means to facilitate the generation of valid and high quality test data for the assessment of chemicals. Expert groups were set up to develop guidelines for the testing of chemicals on one hand and principles of GLP on the other. In 1981, the OECD council adopted a decision concerning the mutual acceptance of data in assessment of chemicals [C(81) 30 (final)].

The first legal measure of EEC level was the adoption of the Council Directive of 18 December 1986 on the harmonization of laws, regulations and administrative provision relating to the application of GLP principles and the verification of their application for tests on chemical substances (87/18/EEC). The second EEC measure came with the adoption of Council Directive of 9th June 1988 on the inspection and verification of GLP (88/320/EEC).

In Japan, the Agricultural Production Bureau of the Ministry of Agriculture, Forestry and Fisheries introduced GLP regulations for pesticides under the Agricultural Chemicals Regulation Law, 1948 in October 1984.

Indian Scenario

Synthetic pesticide usage had a beginning in India with the introduction of DDT for malarial control shortly after the second world war, followed by locust control with HCH (BHC) in 1948 (*3*). After that the usage of pesticides increased in many areas. From the beginning, the importance was given to the effectiveness of pesticides without taking into consideration the toxic effects on their non-target areas before use. The adverse effects caused by improper handling of pesticides during manufacture, transport, storage and sale, and indiscriminate use by the farmers and other users necessitated the regulation of pesticides by registration to prevent risks during manufacture, transport, sale and use. During April and May 1958, many persons died in the states of Kerala and Tamil Nadu (Madras) as a result of food poisoning arising from contamination of food with the organophosphorus insecticide ethyl parathion.

The Government of India appointed the Kerala and Madras food poisoning cases enquiry commission under the chairmanship of Justice J.C. Shah, the Judge of the High Court of Bombay at that time, to inquire into and report on the circumstance under which the food stuff came to be contaminated and the preventive measures to be taken against similar occurrences in the future. The long term measures suggested by the commission envisaged the enactment of legislation to regulate manufacture, sale, storage, transport, distribution and use of pesticides (insecticides, fungicides, herbicides, etc.) in the country. The occurrence of these poisoning cases due to pesticides were not the only ones: there were many cases thereafter. In 1968 an act, "The Insecticides Act, 1968" was accorded the President's assent on September 2, 1968, but became enforceable on October 31, 1971.

During 1978, a subcommittee on pesticide toxicology, chaired by Dr. B.B. Gaitonde, which was constituted by the Registration Committee of the Central Insecticides Board, Ministry of Agriculture, New Delhi, finalized guidelines for data requirements for registration of pesticides in India and protocols for generating toxicity and chemistry data for pesticide registration. The data were generated by Universities, National Laboratories, and reputed laboratories like the Jai Research Foundation, the Rallis Research Centre, and the Fredrick Institute of Plant Protection and Toxicology.

Necessity for GLP Laboratories in India

In 1991, the EEC Council Directive 91/414/EEC was issued concerning the placement of plant protection products on market, and required that the member states could not authorize a plant protection product unless certain requirements were satisfied in accordance with uniform principles provided for in Annex VI of that document. The aim of the directive was to establish and harmonize the data requirements in all the EC member states with respect to authorization of plant protection products. About 89 active substances were scheduled for review under commission regulation EEC No. 3600/92. Indian industries, having shown capabilities for process development for various pesticide molecules, posed a threat to international manufacturers on quality and price. However, they faced problems for export as they had to invest heavily in order to generate data in compliance with Good Laboratory Practices (GLP) from laboratories abroad for registration purposes. Further, due to recent export policies, apart from the agrochemical and pharmaceutical industries, even the dye manufacturers are in need of laboratories certified to GLP. The need for Good Laboratory Practices (GLP) in India was thus felt.

GLP in India

The concept of GLP in India was initiated during 1983 and published in the official gazette. This concept is also similar to OECD principles of GLP, giving scope of definition, testing facilities, management, personnel, QA program, SOPs, facilities for handling test substance and their control, test system, equipment, maintenance and calibration of equipment. However, no major progress was made toward implementing this concept (*4*).

During November 1994, a workshop on GLP was organized by the Hindustan Levers Limited and the Council of Scientific and Industrial Research, Government of India at New Delhi. The workshop recommended the establishment of an Indian GLP monitoring and certifying agency. However, there was no progress.

During December 1995, a one day workshop on GLP (Good Laboratory Practices - an Introduction and Update) was organized at Jai Research Foundation (JRF) where scientists, industrialists and academia from all over India participated. On March 22, 1997, a program was held at JRF when Nigel J. Dent of Country Consultancy, U.K., and Dr. Theo Helder of the Ministry of Health, Welfare and Sport, the Netherlands, gave a talk on Quality Assurance and GLP monitoring, respectively.

On February 8, 1997, a seminar on GLP was organized by the Fredrick Institute of Plant Protection and Toxicology (Fippat), Padappai, which was chaired by Dr. R.L. Rajak, Plant Protection Advisor to the Government of India, and Mrs. Rose Brookes, Quality Assurance Manager, CSL, Hutton, York, delivered the key note address. It was resolved that the Director of the Central Insecticides Laboratory and his team would take the initiative for creating a national GLP inspection authority.

On August 30, 1997, Fippat organized another meeting in which Dr. David Moore, Chief of the GLP Monitoring Authority, U.K., delivered a key note address, and experts from the Department of Science and Technology (DST), Central Insecticides Laboratory and Indian Council of Medical Research participated. The meeting recommended creation of a national GLP Monitoring Authority under the DST and suggested a delegation to meet the Secretary, DST, in this regard.

During the late 1980s, the Department of Science and Technology, Government of India, set up the National Coordination of Testing and Calibrating Facilities (NCTCF). Thereafter, in 1992, it was felt necessary to align the Indian Laboratory Accreditation program to international norms. In the same year, the National Accreditation Board for Testing and Calibration Laboratories (NABL) in the place of NCTCF was formed, which is in line with ISO/IEC Guide 25 (1990) and also aims to bring the criteria in line with requirements of Europe as in EN 45001 (1989). *NABL* maintains its linkages with the international bodies like International Laboratory Accreditation Conference (ILAC) and the Asia Pacific Laboratory Accreditation Co-operation (APLAC) by participation in their conferences. The testing and calibration in India are being accreditated by NABL at present. Though private laboratories are being accredited by NABL, it is envisaged in the near future that national laboratories may also be required to fall in line with NABL accreditation.

GLP Facilities in India

In India, very few laboratories, except the Jai Research Foundation (received GLP certification in 1996 from the Veterinary Public Health Inspectorate, GLP Section, pursuant to the Netherlands GLP Compliance Monitoring Programme, Ministry of Health, Welfare and Sport, State Supervisory Public Health Service, Government of

The Netherlands) and the Rallis Research Centre (received GLP certification in 1992 from the German Health Ministry, BGVV) conduct studies in accordance with international GLP principles.

Contract Services. Today, the facilities compliant with GLP in India can offer contract services on the following:

- Bioefficacy including phytotoxicity and compatibility meeting Good Field Practices
- Physico-chemical analysis
- Residue studies
- Environmental fate studies
- Metabolism studies
- Toxicology including acute, subchronic and chronic studies; genetic and reproductive toxicology
- Ecotoxicology
- Biomonitoring exposure of workers

Third Party Laboratory Accreditation. Quality Assurance Services, Australia, an independent body, has certified three laboratories in India *viz.*, Jai Research Foundation, Vimta labs. Ltd. (for toxicology and chemistry) and Gharda Chemicals Ltd. (for chemistry) as complying with the requirements of GLP. The implications, status and significance of such third party accreditation in relation to international acceptance of the study data is not fully understood.

Problems and Challenges

The problems and challenges one has to face in implementing GLPs in India are outlined briefly.

GLP Certification and Monitoring. At present no GLP certification and monitoring system exist in India. However, recently the Adviser from The Department of Science and Technology (DST) has been impressed on the need to create a GLP monitoring Authority in India and hopefully it should take shape soon.

In India, the registration of pesticides comes under the Ministry of Agriculture, and the guidelines for data generation have been laid out by the B.B. Gaitonde committee appointed by the Ministry of Agriculture. In order to have a GLP monitoring system implemented in India it is essential that the existing protocols defined by the Dr. B.B. Gaitonde committee and approved by the Ministry of Agriculture be required to be statutorily revised to fall in line with the international guidelines. Recently, the Ministry of Agriculture has constituted committees to address these concerns.

The Indian laboratories have to cope with certain difficulties, such as, power cuts; irregular supply of basic items like quality animal feed supported by feed analysis reports; obtaining the services of Veterinary pathologists experienced in rodent pathology; difficulty in getting animals of required age and weight within short time frames from the Indian suppliers (a herculean task to get an import license for the animals from abroad); difficulty in getting certain instruments calibrated as calibration facilities are limited in India; problems faced in explaining to Indian sponsors the time and investment needed to generate data in compliance with GLP; sometimes adequate basic information on test substances is not provided by the sponsors and must be obtained before commencement of the study; and delays in obtaining the permits from the Central Insecticides Board for importing samples from abroad.

Prospects in India

Many Indian companies will have the opportunity to utilize the facilities and expand their export market. Being competitive in price will attract a considerable amount of contract work from sponsors from different countries whose cost of data generation is highly prohibitive or who do not have laboratories that comply with the GLP standards. The capability of providing the complete acute toxicology studies and the physico-chemical and analytical chemistry services would make contract facilities in India indispensable, competitive and attractive. Many pharmaceutical R&D labs also have started implementing GLP programs for in-house evaluations of new drugs. The Government of India, in the near future, will probably insist that data generated in Indian laboratories follow GLPs. Under these circumstances, the GLP facilities will become indispensable in the country. The Government of India soon will be establishing an Inspectorate for monitoring and certifying laboratories for GLP compliance.

Acknowledgments

Thanks are due to Mr. R.D. Shroff, President, and Mrs. Sandra R. Shroff, Vice-President, JRF, for encouragement and support in the establishment of a GLP program in JRF and for providing the opportunity to attend the conference at Cancun, Mexico.

Literature Cited

1. *Good Laboratory and Clinical Practices;* Carson, P.A; Dent, N.J., Eds.; *Heinemann Newnes*, 1990, pp 390.
2. Broad, R.D; Dent, N.J. An introduction to good laboratory practice (GLP) in *Good Laboratory and Clinical Practices;* Carson and Dent, Eds.; *Heinemann Newnes;* 1990; p 3-15.
3. David, B.V. Pesticide Industry in India. In *Indian Pesticide Industry*, David, B.V., Ed.; Vishvas Publications, Madras, 1981, pp 1-36.
4. Kanungo, D. Good Laboratory Practices (GLP) - an Introduction and Update -training programme at JRF on 4th December 1995 - Inaugural address. *Indian J. Environ. & Toxicol.;* 5(2): pp 49-50.

Chapter 33

Validation of Complete GC and HPLC Systems in Analytical Chemistry: Is Validation of Individual System Components Really Necessary?

Michael Williams

Horizon Laboratories, Inc., 1610 Business Loop 70 West, Columbia, MO 65202

The performance of GC and HPLC systems must be validated by the analytical chemist. Some investigators debate that components of these systems must be validated individually; others contend that systems can be validated as intact units. This paper will argue that through use of certified calibration standard(s), an exhaustive review of the data, and proper study documentation, complete GC and HPLC systems actually validate themselves during their use.

In today's modern laboratory, typical gas chromatography (GC) and high performance liquid chromatography (HPLC) systems are fully automated. Following a period of system set up during which operational parameters are identified, samples can be analyzed successfully with little to no intervention by the user. Such technology has enhanced laboratory throughput and efficiency because time management is improved; more samples per unit time can be prepared in the wet laboratory, then analyzed unattended throughout the day and night hours. Precision and accuracy are also enhanced; current GC and HPLC systems are manufactured to high tooling and electronic specifications and, when they function properly, their precision and accuracy exceeds human performance.

But analytical chemists pay a severe price for unattended automation. In our absence, we lose valuable, real-time information about instrument performance. Before the birth of autosamplers and data collection devices, instruments were closely tended by chemists, if for no other reason than samples were hand injected and certain extracts required dilution when peaks climbed off scale. Problems were immediately evident and corrective action taken, in many cases without sacrificing the run. With automation, chemists arrive at the laboratory with bated breath and confront a mass of data in the form of paper or bytes, and must judge whether all went well, or all went wrong, in their absence. If things went well, we deem automation as wise, efficient, and effective; if not, we recite words best not printed here.

Automation absolutely requires that ALL components of very complex GC and HPLC systems work properly, especially when analyses are unattended and problems cannot be detected immediately. In spite of their sophistication, GC and HPLC systems do break down or, more likely, their performance degrades over time in response to a variety of factors related to sample matrix, clean-up procedures, impure gases, extended use, etc. Hence, a central question to analytical chemists, their management, study sponsors, and the regulatory community is "During a study, and at the end of a study, how do we validate complex GC and HPLC systems and the data these systems generate?"

To be sure, these systems do need validation. The configurations of modern GC and HPLC systems are quite complex and vary considerably depending upon the particular task at hand. Typical components can include the following, in sequential order: (i) autosampler, (ii) injector, (iii) analytical column housed in an oven (GC) or at near-ambient temperatures (HPLC), (iv) detector, (v) electronics plus attendant integration algorithm(s), (vi) electronic data storage, and (vii) a printer for "hard" data. Each is a potential weak link that can compromise the integrity of the resulting data: autosamplers inject the wrong sample, injector needles/transfer lines clog, glass liners become contaminated, columns degrade, temperatures vary, detector sensitivities drift, and baselines are improperly drawn. Any one of these or hundreds of other potential disasters can destroy an otherwise fine analysis and drive a harried chemist to whimpering despair.

How can we be sure these instruments have worked properly?

"Absolute" Versus "Consistent" System Performance

There are two major philosophies regarding how to validate GC and HPLC systems: (i) measurement of "absolute" system performance, and (ii) measurement of "consistent" system performance. Proponents of the "absolute" technique believe that validation of complete GC or HPLC systems is best achieved by periodically returning individual system components to their original factory specifications or some other standard condition, then documenting that fact as a part of the component's maintenance log. For example, autosamplers should be verified independently for predictable sequencing of samples, injection volumes, needle residence times, and effectiveness of the system wash cycle. Similarly, GC column ovens require verification of absolute temperature and program ramp rates. Columns must be calibrated by a "standard" technique. Detectors and their associated output signals must be adjusted to standard specifications, and so on. Such actions are generally followed by re-assembly of the intact system, then injection of factory-prepared solutions under standard conditions, yielding a chromatogram(s) which must possess certain standard attributes. If this process is completed in a satisfactory manner, the system is considered validated for future use.

This approach is certainly rigorous and does possess one notable merit; the strategy yields baseline performance norms against which all that follows can be measured. Factory settings are often modified and made project specific, and there is a certain comfort in knowing from whence you came. But the most telling criticism of this technique is that it gives information only from a single point in time when the component is deliberately removed from operation and placed in factory or standard condition. Such information says very little about the quality of data obtained AFTER the component is returned to operational status. Indeed, such information provides no direct evidence of ongoing or future instrument performance; reviewers of the work can be grossly misled that all was well when in fact it was not, a conclusion based solely upon the comforting knowledge that an "absolute" system performance measurement was made in the past. The very act of returning an instrument to operational status obviates the entire "absolute" validation process relative to future performance if project-specific modifications are made. All that can be concluded with any degree of certainty is that the component/system performed properly BEFORE it was made project specific. Nothing can be said about AFTER, which is where the real data are generated.

Measures of "absolute" system performance do have real value when (i) a new unit arrives on the loading dock, (ii) something goes seriously awry that cannot be easily discerned or repaired, (iii) a rigorous, back-to-the-basics, preventative maintenance program is desired, and/or (iv) a method is transferred from one laboratory to another. Purchasers of expensive instrumentation certainly want their units in prime condition upon receipt, and a complete system evaluation is prudent and warranted before the unit enters operational status. When instrument components seriously malfunction, quite often the best recourse is to rehabilitate the unit to factory specifications, then proceed with project-specific modifications from there. "Absolute" system performance measurements also make for a fine, albeit expensive, preventative maintenance program, but there is a concomitant risk that units will not easily return to their previous operational status if the analytical separation is sensitive and difficult. Accurate instrument specifications are useful when a method is transferred from one laboratory to another, but the value of this information is mitigated since lab-to-lab variations in instrumentation, columns, detectors, personnel, etc., can be sizeable; experienced chromatographers use the proffered instrument specifications only as a starting point, then modify with the goal of accomplishing the task, not reproducing someone else's work exactly.

The most pertinent information about how an automated system has performed is obtained AFTER an analytical sequence is complete. This valuable information is contained in the raw data assuming, of course, that the analysis was properly documented in the first place. These raw data consist of chromatography, print outs of instrument conditions, and documentation of pre- and post-run activities. Additional, and absolutely essential, information is contained in documentation of the calibration standard used to quantify the analyte(s) of interest. By judicious, informed, and logical examination of these data, accurate and irrefutable evidence is obtained which validates (or invalidates) the performance of the entire instrument system and its individual components **during their operation**.

The logic behind this conclusion is that accurate and precise analytical data depend mainly upon the "consistent" performance of GC or HPLC instrumentation. An instrument system may not be at optimal or factory specifications, but if it performs the same way, every time, for the duration of an analytical sequence, and the underlying conditions yield unambiguous chromatography, then accurate and precise measurements will be made. For example, it does not matter whether a GC oven temperature ramps at 5°C/minute or 7°C/minute, so long as it behaves reproducibly time and time again. Similarly, an HPLC injector assembly may deliver 55 microliters rather than 50 microliters each time, but as long as it delivers 55 microliters for samples and calibration standards alike, it will yield accurate and precise analytical measurements. Knowledge of "absolute" instrument parameters is seldom a required condition for judgement of accuracy and precision; however, "consistent" performance is an absolute requirement.

Of course, no chromatographer should use any instrument which is badly out of calibration or on the brink of serious malfunction. Such acts are inefficient, a misuse of fine instrumentation, and represent shoddy, unprincipled work. For example, a detector which performs at 1/10th its normal sensitivity should be repaired or replaced, even if it behaves consistently and meets project limit of detection specifications. Such performance is an indication of imminent failure.

Upon completion of a properly documented analytical sequence, the following information should be available to the analytical chemist:

- Sample and calibration standard order of injection,
- Chromatography (bytes and/or paper) from calibration standards, untreated controls, fortified controls, and authentic samples,
- Peak height/area measurements,
- Printouts of instrument parameters (i.e., temperatures, gas/liquid composition and flows, data collection settings, injection volumes, etc.), and
- Documentation of calibration standard identity and purity.

These data contain all of the evidence necessary to validate the instrument system and its components by a "consistent" performance measurement. That is, the entire system, and each component of the system, places an indelible stamp of proper or improper performance directly into the raw data. In this paper, this evidence is termed "Key Indicators"; the component features which require validation are termed "Critical Components". A typical argument for GC research is presented below; the argument for HPLC proceeds similarly, but is not presented here.

Autosamplers

Autosampler devices are robots which sequentially inject sample extracts into a GC in a systematic, predetermined manner. The critical components of autosamplers that must be validated are (i) injection order, (ii) injection volume, (iii) needle residence time, and (iv) efficacy of the between-sample wash cycle. The key indicators for autosampler performance are the calibration chromatograms and the associated calibration curve(s).

In any well-designed analysis, solutions of calibration standards are interspersed with authentic sample extracts; sequences also begin and end with calibration standard injections. The projected order of injection is documented before the analysis. The chemist eventually correlates the anticipated injection sequence with chromatograms obtained during the analysis, identifies the calibration chromatograms, and constructs the appropriate calibration curve(s).

A calibration curve prepared from three or more concentrations of the calibration analyte and possessive of an appropriate concentration versus peak height/area response can be constructed **if and only if** the injection order was as anticipated. If the autosampler did not inject the samples/calibration standards in the proper order, sample chromatograms will be confused with calibration chromatograms and the calibration curve will bear no resemblance whatsoever to what was anticipated. Similarly, variable injection volumes and needle residence times will yield poor concentration versus response correlations since variable amounts of analyte will have been injected. The very fact that a proper calibration curve can be constructed is *prima facie* evidence that critical components of an autosampler performed consistently throughout the sequence; it is simply not possible to construct a proper calibration curve if one or more of these critical components is awry.

Unacceptable cleansing of the injection needle and transfer lines will be manifested as spurious sample matrix peaks in calibration chromatograms injected after authentic samples; the peaks will not be present in calibration standards injected at the start of the run. The absence of spurious peaks is excellent evidence that the between-sample wash cycle was efficacious.

Injection Ports

The injection port of a GC volatilizes the calibration standard/sample extract, then loads the resulting gas onto an analytical column. Injector configurations are highly variable and project specific; they may include glass liners of various types operated in split, split-splitless, or splitless modes. In all cases, the critical components that must be validated are (i) intactness (i.e., no gas leaks), (ii) gas flows (septum purge and split ratio), (iii) split-splitless timed events, (iv) injector temperature, and (v) presence or absence of analyte discrimination. The key indicators in the raw data are (i) the calibration curve model and (ii) chromatography and calculations from quality control samples analyzed parallel with the samples.

The calibration curve model plus the associated chromatograms yield excellent information about the performance status of the injection port. It is simply not possible to obtain acceptable chromatography and a satisfactory calibration curve if the injector has a gas leak, gas flows are unstable, or split/splitless timed events are variable. Gas leaks will be manifested in the chromatograms as wider-than-normal solvent fronts. If the leak varies from injection to injection, analyte retention times will vary and peak shapes will be askew. Unstable injector gas flows primarily affect chromatography from instruments operated in the split mode; peak heights, widths, and retention times from calibration standards will vary and yield poor calibration curve correlations. Poor calibration curve correlations will also result if split/splitless timed events or actions are inconsistent since variable amounts of analyte will load onto the analytical column.

Variable injector temperatures affect volatility of the analytes and other substances present in the sample, especially if the temperature fluctuation encompasses the boiling point of the analyte. Peak heights/areas of calibration standards will vary and poor calibration curve correlations will be the result. In addition, injection ports can show discrimination for or against analytes of interest depending upon the cleanliness of the sample and concentration of analytes and other matrix substances in the final extract. Final extracts from untreated control samples and controls fortified at the limit of quantification (LOQ) contain the highest concentration of extraneous matrix material since the final volume of these extracts necessarily must be small (i.e., to measure small LOQ peaks). High concentrations of matrix materials can bias analyte recoveries high or low; hence, these samples provide evidence for or against such injector bias. Similarly, high level fortifications give information about injector performance when the concentration of matrix materials is low since these samples are generally diluted to fit the range of the calibration curve. If analyte recoveries are similar between LOQ and high-level fortifications, and it is known that sample manipulations prior to GC analysis show no concentration bias, this is excellent evidence that matrix components have no deleterious effects on the analyte as it passes through the injector.

If proper calibration curves can be constructed, and no bias is observed by comparison of recoveries of LOQ and higher level fortifications, the injector has validated itself through its demonstrated performance. These data are *prima facie* evidence that the critical components of the injector performed consistently throughout the sequence.

Column/Oven

The analytical column separates analytes of interest from interfering components so the former can be measured by the detector. In gas chromatography, the column is heated in an oven so as to maintain the gaseous nature of the incoming extract. The critical components of the column that must be validated as working consistently are (i) overall column performance in effecting the separation, (ii) oven temperature(s), and (iii) column carrier gas flow. The key indicators are (i) the calibration curve model, (ii) quality control sample chromatography and recovery results, and (iii) the characteristics of sample, calibration standard, control, and fortified-control chromatography.

The performance of analytical columns can be profoundly influenced by the presence or absence of matrix components. These interferences can gradually degrade the resolving power of the column and affect analyte migration. Chromatography from calibration standards is an excellent measure of column performance in the absence of matrix; if the chromatography indicates reproducible retention times, acceptable peak shapes and sensitivity, and no evidence of analyte degradation, then the column is validated with respect to consistent chromatography of calibration standards. Similarly, quality control samples provide a measure of column performance in the presence of matrix. LOQ fortifications test the column under conditions of high matrix loads; high-level fortifications perform the same test under conditions of low matrix loads. So long as peak shapes, sensitivities, resolution from interferences, retention times, etc., do not change throughout an analytical sequence between calibration standard and fortified-control chromatography, the column remains validated by its demonstrated consistent performance.

A general examination of all chromatography will reveal much about column performance. If retention times shift during the run, column performance may have degraded or gas flows may be fluctuating. Peak shapes should remain constant throughout a run, even in the presence of matrix. Development of peak asymmetry, loss of resolution between the analyte(s) and interferences, or baseline drift during the sequence may indicate column degradation or overloading and concomitant loss of validated status.

If proper calibration curves can be constructed, retention times and peak shapes do not vary with time or matrix, peak resolution remains constant regardless of sample load, and no bias is observed in a comparison of LOQ and higher level fortifications, the column has validated itself through its own performance. As with the previously discussed components, these data are *prima facie* evidence that the critical components of the analytical column performed consistently.

Detector

The detector measures the analyte as it exits the analytical column. The sophistication of current GC detectors is remarkable, ranging from the relatively simple flame ionization detector to highly selective and complex mass spectrographs. The critical components which must be validated are (i) sensitivity, (ii) selectivity, (iii) gas flows, and (iv) temperature/voltage. The key indicators are (i) the calibration curve model, (ii) quality control sample chromatograms and results, and (iii) the general condition of the chromatography.

It is simply not possible to construct an acceptable calibration curve if detector sensitivity towards an analyte varies during an analytical sequence; calibration standard responses will vary with time and will confound construction of the curve. Similarly, detector temperature/voltage and gas flows have a profound effect on the sensitivity of detectors and, if these parameters fluctuate, construction of a proper calibration curve is not feasible. Construction of a scientifically logical response curve from a set of calibration chromatograms is *prima facie* evidence for consistent detector response in the absence of sample matrix.

Chromatograms from quality control samples yield evidence for detector performance in the presence of matrix. For example, final extracts from LOQ fortifications contain the highest concentration of matrix interferences, and while these interferences may not be detected (due to detector selectivity), they may influence detector responses to an analyte. Such situations can occur when nondetectable interferences (e.g., phthalates, elemental sulfur, PCBs, etc.) co-elute with, or elute near, an analyte, temporarily modifying a detector's sensitivity. Satisfactory chromatography and recovery of LOQ analytes are excellent evidence that detector sensitivity and selectivity have not been compromised by high matrix loads. A similar argument applies to low matrix loads from controls fortified at high levels.

Much evidence of detector performance may be obtained through examination of the general condition of all chromatography from an analytical set. The absence of voltage "spikes" and negative peaks, constant background currents, reproducible calibration standard and sample responses, absence of peak asymmetry, acceptable signal-to-noise ratios, stable baselines during the run, etc., are all evidence for constant and acceptable detector performance. Taken *in toto*, the key indicators described above validate, or invalidate, detector performance.

Data Collection Devices/Integration Algorithms

Perhaps no other component of GC systems has received as much validation scrutiny as data collection devices and their attendant integration algorithms. This is especially true for systems which allow user access prior to the printing of final chromatograms and calculated data. There is much justification for this attention; user access/intervention is considered proven, fruitful ground for mischief and fraudulent manipulation of results.

Data collection systems collect raw electronic signals from GC detectors. Older model integrators and recorders process these data into peaks superimposed upon a baseline; integrators use pre-set baseline rules and algorithms to calculate peak heights/areas, but the user must manually make these same measurements with recorders. In either case, there is little opportunity for user intervention with these types of units. Validation of these instruments has generally involved the periodic use of peak generators which calibrate the linearity and reproducability of the device.

Newer data collection systems are much more sophisticated. These units use computers and data storage devices which allow retrieval and manipulation of raw data hours, days, weeks, and even years after the data have been collected. Manufacturers, laboratory managers, chemists, and the regulatory community quickly recognized that such systems must be protected against unauthorized, unrecognized manipulation. For the most part such protections are now in place, but knowledgeable, unethical "hackers" still pose a formidable challenge.

User intervention is certainly a desirable feature of these systems. After all, each chromatogram is unique and full of challenge for even the most powerful of computers and algorithms. The fact is, such systems cannot yet match the judgement of highly trained, knowledgeable chromatographers. Interfering matrix peaks, baseline perturbations, subtle changes in analyte retention times, etc., can mislead a computer and yield inaccurate, imprecise information. Such mistakes can only be corrected by the user; the computer simply does not know how.

The trick is to manage such systems so that such manipulations are either prevented entirely, or identified by user name and date as they occur, as appropriate. Example activities which must always be prevented include (i) alterations of original electronic raw data signals and (ii) manual entry of an integration value after peak integration has occurred. Permitted interventions embrace the correction of baselines, peak start and stop points, peak skimming inaccuracies, merged peak droplines, etc., so that proper integration can occur. With proper preventions in place, modern data collection systems have provided the chemist with powerful tools that facilitate efficiency in the modern laboratory.

Given that proper standard operating procedures and diligence are in place to document user intervention, how do we validate data collection devices and their attendant algorithms? Many GCs have internal programs which, when initiated, send a series of electronic signals to the data collector. These tests measure system linearity and integration consistency; the results should conform to certain "standard" specifications that validate accuracy and precision. This type of "absolute" system measurement is quick, easily documented, and does not require dismantling of the collection device from the main unit. Use of an independent peak generator which must be deliberately connected to the data system has similar virtues.

However, as with all other GC system components discussed previously, this approach is grossly deficient in one key way: It measures performance at a single point in time when "real" data are not being collected. Just because the data collection system performed correctly during an "absolute" system test does not mean that it did so during actual use. Fortunately, there are key indicators in the data that validate the data system during actual use. They are: (i) the calibration curve model, (ii) quality control samples, and (iii), most importantly, the analytical calibration standard which has been characterized according to Good Laboratory Practice standards. The critical components of the data system that must be validated are (i) the electrical connections (cables, chips, connectors, etc.), (ii) the integration algorithm, and (iii) the mathematics.

The calibration curve validates the electrical connections and integration algorithm in the absence of matrix. It is simply not possible to construct a valid calibration curve if either of these critical components are malfunctioning or variable with time. The use of a properly characterized calibration standard that has been accurately weighed and diluted is critical to this evaluation; the calibration curve series serves the same function as the electronic peak generator described above except that it measures the performance of the data capture system **during the generation of authentic data with the certified analyte of interest**. If the calibration curve is scientifically acceptable, then the data capture system has performed consistently throughout the analytical sequence. Further, if the calibration standard has been prepared and characterized properly, then accurate and precise measurements will be made even if the data capture system performs consistently, but not at factory specifications.

Quality control samples serve a similar function for samples containing matrix. Matrix is seldom completely removed by clean-up techniques from final extracts. As a result, extracts may contain interferences which elute near the analyte(s) of interest. A careful evaluation of the associated chromatography from these samples, with particular attention given to peak start and stop marks, drop lines, and peak shapes, help the chromatographer judge if the data capture system has integrated the peaks consistently and correctly.

Most data capture systems can also perform additional mathematical calculations on the data. Generally, these manipulations include calculation of a variety of least square calibration curve models plus subsequent calculation of residue levels in unknown extracts and their originating samples. These calculations mostly use simple mathematics which can be checked by a hand-held calculator. Or, they can be checked by hand if one believes calculators should be validated also. A more common approach is to load the computer program with a "dummy" set of data for which the mathematical solutions are known and validated. Either way, the mathematics of the data capture device can be validated in simple fashion.

Summary and Conclusions

Analytical chemists have a huge responsibility. We must provide accurate and precise data so that important business and policy decisions can be correctly made. With that responsibility comes the obligation of defending our data to others, proving that the research was performed competently with no errors beyond normal scientific variation. How well we accomplish this defense depends upon the soundness of our science plus our ability to document that science.

Many factors are involved in such a defense; this paper addresses only one: Validation of the instruments that generate the final data. This is important because if a chemist cannot demonstrate that instruments were operating in a scientifically acceptable manner, how can he/she attest to the accuracy and precision of the final data?

The author believes that the integrity of data from GC and HPLC systems can only be determined by a comprehensive evaluation of key chromatographic indicators and other ancillary raw data obtained during a study. These key indicators validate, or invalidate, the consistent performance of the complete system and each system component. The reader will note that system validation can only be inferred when ALL key indicators in the raw data meet generally accepted scientific norms. If any one key indicator fails to meet these norms, then the instrument system should be considered invalid pending correction of the underlying problem. Individual key indicators do not necessarily identify the component responsible for invalidation of a system; as noted above, specific problems with a given key indicator can have multiple origins. Even so, highly-trained chromatographers generally possess sufficient knowledge, insight, and intuition to narrow the field of possibilities.

For validation purposes, measures of "absolute" instrument performance have limited merit except, as discussed, in the most narrow of circumstances. Indeed, such validations can provoke a false sense of security. Such measures provide no pertinent information about ongoing or future system performance; "constant" measurements, however, evaluate very real, ongoing events that occur during generation of the desired analytical data.

Measurements of "absolute" system performance are <u>not</u> an effective substitute for the comprehensive training of chemists, management, sponsor, or regulator. Such training is an absolute requirement for making "constant" performance measurements and concomitant validation of these systems. There are NO effective shortcuts for anyone.

Chapter 34

Validation, Vertification: Possibility, Probability

Richard E. Cooney

**Office of Enforcement and Compliance Assistance,
U.S. Environmental Protection Agency, Washington, DC 20460**

"Behold, and learn to do justice and contemn not the gods!"
The Aeneid, Virgil, 31BC

It is not the intent of the author to present a procedural tutorial on how to do a software verification or system validation process, but rather to evolve the processes of system engineering that employs the need for verification and validation and thus obtain for the system user an appreciation of what the process entails. This premise is based on experience obtained from the inspection of numerous scientific facilities throughout the United States. The inspections revealed that with one notable exception none of those inspected possessed the necessary facilities, expertise and personnel to conduct a qualified verification and validation procedure. The process of verification and validation requires extensive education and experience and the idea that it can be both presented in a short monograph such as this or conducted in a manner similar to balance weight checking is without credence.

The requirements for software verification and systems validation are not specifically addressed in any rules associated with the Good Laboratory Practice Standards. Reference is made to 40 CFR §160.61 and 40 CFR §160.63 that equipment, "must be of appropriate design and adequate capacity to function according to the protocol or SOPs," and "adequately inspected, maintained and calibrated," respectively. As a result of this very generic statement there has been a plethora of schemes, plans, and programs developed, within the GLP regulated environment, for the conduct of verification and validation of computer systems. Everyone has the ultimate solution or better still, **their** ultimate solution is the only possible one. History and tradition may not support a solution for, "In spite of the extensive research effort by government, industry, and academia, no single methodology has surfaced to eliminate these unwanted trademarks" *(1)*. The trademarks referred to here are software errors, bugs and glitches.

Why is it that with the many varied types of equipment available to a laboratory the stand alone computer system or reporting computer is the only one culled out to be validated and verified? The balance is determined to be accurate by using registered weights for comparison; chromatography uses a blank; substance is characterized; temperature and humidity measured and numerical computations are checked for accuracy. No other laboratory device, either stand alone or grouped, is required to be validated and verified. It is my personal opinion that the terms should be dropped and replaced with the singular term, either audit or calibrate. The audit process is well known in the computer world. Security audits a system to insure that no one is causing harm to the system, either intentionally or unintentionally. Management audits a system to insure accuracy and proper use. Diagnostics audit the system to attest its functional, electrical and mechanical properties. None of these functions satisfy the full process of verification and validation.

The underlying intent of this briefing is to open meaningful dialogue, within the regulated community, on the reality of verification and validation. Therefore, the title *Validation, Verification - Probability, Possibility*, should act as an initial catalyst for such a dialogue. The need for the Validation and Verification (V&V) process is very real, but the probability of a laboratory conducting a complete and successful one is not only remote but inappropriate. The possibility of a laboratory to conduct a validation and verification process is only attainable if the person(s) conducting such an endeavor are properly trained and have acquired the necessary expertise through education and experience and the facility has the necessary tools to do V&V. The process cannot be accomplished without access to the basic items of source code and hardware configuration. V&V are finite processes that cannot be left to individual interpretation nor can they be conducted as an adjunct to some other task. Most software and hardware have already undergone V&V by the developer and builder. Why try to do that which is not only unnecessary but could cause serious problems with installed systems.

An understanding of the terms as proffered by the National Computer Security Center (NCSC) is an excellent point of departure for discussion. "Verification is the process of comparing two levels of system specification for proper correspondence, e.g., top-level specification with source code, or source code with object code. This process may or may not be automated" and, "Validation is the process of insuring that the software, hardware, and connectivity is proper, adequate and capable of doing the assigned task. This process may or may not be automated." NCSC is a function of the National Institutes of Standards, and Technology, formerly known as the National Bureau of Standards (NBS), and the National Security Agency (NSA). The mission of the NCSC is to establish the policy by which all U.S. Government systems are developed and produced. System in this sense is not element specific but includes systems, software, firmware, middleware as both independent elements and as a synergism. ***Verification applies to a process while validation applies to a system***. The United States Department of Defense standard, DOD-STD-2168 states, "Independent Verification and Validation (IV&V) - Verification and validation performed by an organization that is both technically and managerially separate from the organization responsible for developing the product or performing the activity being evaluated." The standard prevents the user from performing the IV&V and in the GLP environment that would mean the laboratory. The logic for stating a DOD standard is that for most of the national standards for information processing have their genesis in the DOD. The federal standards are referred to as Federal Information Processing Standard(s) (FIPS). The FIPS series establishes the benchmarks by which all systems are tested in order for them to be considered buyable by the government. By default FIPS has become the industry standard so the translation from DOD to industry is almost transparent. Like it or not we all live by them - Energy Star is an example of a government developed criterion which industry uses.

Basic to all systems/software development is the process historically referred to as Life Cycle Development. Figure 1 displays this process. The cycle has been annotated for software development but the same type of cycle exists for hardware, firmware and middleware. I know of no one who has attempted to validate the hardware or verify the DOS, Windows or for that matter the most rudimentary part of the computer - the BIOS or Basic Input Output System. The cycle is circular to demonstrate the continuing activity in the development cycle. Giving credence to this practice, How many of you are using the initial versions of DOS, Windows, WordPerfect, Microsoft® Word, Excel? To use a phrase not of my origin but of a source I know not, "Technology proliferates as though it were born pregnant." While not as proliferate but quite active is hardware. Software, hardware, firmware and middleware share a dynamic obsolescence cycle. Some may argue that the obsolescence is planned so as to create growth in the industry while other will say it a common result when a basic idea is defined and then refined by lessons learned. Whatever the reason, the growth of the personal computer associated business has been meteoric. The cycle has a definite, definitive step named Independent Validation and Verification (IV&V). It is an integral part of the cycle and cannot be neglected or passed over. The cycle continues even after the product is released for consumers for how else would "Upgrades" become available?

The process of Verification and Validation requires an intimate knowledge of the hardware, firmware and software. The knowledge is intuitive in that as a system is developed and brought into production the software and system matures within the development cycle. Progress through the cycle, with the exception of documentation, is based on successful producing of a useable item with or based on the previous step. The item may be retained, changed or modified at a later point.

There are generally two methods of V&V. One is as a function of the developmental cycle and the other is done after the system has attained pre-delivery status. Summation of the two - proactive and after the fact. The proactive application is usually done on systems that are developed in house where the V&V team is an integral member of the developmental cycle group thereby acquiring instinctive knowledge of the software/system as it matures. The proactive group, siting in on all phases of the development cycle, obtains the widest possible exposure to the software system in development. The proactive group is by far the most knowledgeable of the system and to express a parallel, functions as the GLP Quality Assurance Unit for the development process. "Formal verification is the application, in a rigid and algorithmic fashion, of mathematical and logical principles to the problem or certifying the computer programs are correct and consistent" *(2)*. Another definition follows: "Verification is proving algorithms correct ... and, when more than one algorithm is known to solve a problem, a comparison of their relative efficiencies" *(3)*. This group is responsible for the verification aspect.

Figure 1. Development Life Cycle - Software

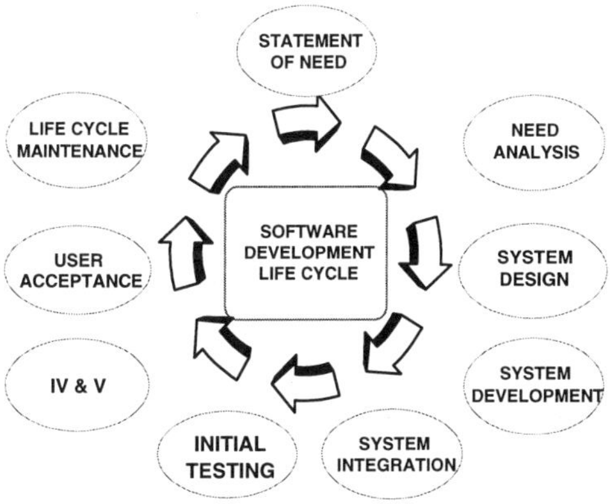

The after the fact group integrates into the process at individual segment completion compiling data to be used for overall V&V and final system validation. This group usually records the statement of need and all modifications thereto, all design changes, functional data descriptions and data elements, change proposals, and configuration management aspects. These data are used when an overall system V&V is done. This is more closely aligned with systems validation process and satisfies the desired objectivity by being independent of the developers thus, IV&V. The IV&V team will review and sometime conduct their own verification of the systems associated developed software. For purposes of this discussion I will not differentiate not define the differences between IV&V and V&V and the terms will be used synonymously.

Within the cycle some of the functions can be conducted in parallel. Statement of Need, Need Analysis and System Design work in concert with each other as does System Design, System Development and Systems Integration and so forth. One element of the cycle that is not delineated as a separate entity but has overall activity and is understood to be prime is the Documentation component. Documentation is a example of a function that continues from the beginning and never seems to end. Documentation is the 'paper work' associated with the system or the software. The user manual is a function of the Documentation procedure as is the maintenance manual, configuration management, source code, language, logic, change and version records, and alpha and beta testing results. "By comparison with most other intellectual pursuits, in good engineering design a great deal of attention is paid to documentation; indeed, great attention is paid to detail by the traditional engineer no matter how tedious it may seem to others" *(4)*. There is, within the computer development world, a documentation discipline practiced by special type of people who command high salaries for their services. Jesting, being a documentation guru guarantees long term employment for the process never seems to end and without that expertise your product is doomed to failure. Ninety five percent of developmental life cycles and the associated hardware or software that fail prior to successful production are caused by lack of, unsupported or incomplete documentation. V&V must have all available documentation for their task to be feasible.

The development cycle starts with the Statement of Need and the drafters of the statement should have "...a complete: understanding of policies and procedures; understanding of hardware and software; understanding of application area tasks; understanding of the perspectives, needs and capabilities of the users" *(5)*. Is this capability present in the laboratory? Certain 'understandings' are present, but in totality laboratory personnel do not have the intimate knowledge required of these 'understandings'. The progress through the cycle is time driven. A developmental cycle is time constrained and has established time nodes. While it has the appearance of being open-ended, the cycle does have a time to meet for production, publication or prototype. Often the software developers will release a pre-production copy of a product for the sole purpose of acquiring an exposure to the user community and thus obtain a commitment by the user. Changes to the pre-production copy are usually provided free of change. Consider for a moment Microsoft's pre-release of the various iterations of Windows. This is not to be confused with the issuance of a BETA version of software. BETA copies are provided to selected users for evaluation and comment. BETA version usually have a time period of usage an will cease to be usable after the period has expired.

Continued use of a time bounded copy of software, while giving the impression that all is well, can cause havoc, transparent to the user, within the operating system of a computer as well as having an effect on other applications software.

Software has certain factors that must be addressed while in development, production, purchase and use phases and these must be considered during the V&V phase. Figure 2 presents the factors with the expansion of the individual attributes listed below. The expansion provides an excellent list of questions that can be asked by even the most uninitiated. The questions can be asked during or prior to the purchase of application software and to some extent hardware without having to entertain a V&V function. They need not all to be asked nor answered but select those that are appropriate for the circumstance and they will prove to be an aid of value.

As stated previously that as most of the suggested V&V efforts of a laboratory are directed towards applications software. The applications usually are of the support type and not mission. Sciex software is never mentioned as a candidate for verification nor is Datalogger, Hobo or Chemtron. A basic need to do a successful verification process is the source code. I do not know the exact number of lines of executable code associated with WordPerfect but and educated guess would be in excess of one hundred thousand. A more complicated program such as one to do chemical analysis may have in excess of four-hundred thousand lines of executable code. The language, logic, application and interactivity of the system has serious impact on the size of a program. Consider what would be the size of the Microsoft® Office software package containing Word, Excel, Access, Powerpoint and Bookshelf *(8)* programs. The entire program is contained on two CDS and Bookshelf is never loaded but accessed from the CD. In the past, prior to CDS, Windows 95 had 13 3.5" discs, Power Point, 10 3.5" discs, Access, 14 3.5" discs. Would anyone care to estimate the number of lines of source code associated with Microsoft® Office. Sheer size alone should dissuade anyone from attempting to verify the suite. To do verification you must evaluate the software and without the source code this feat cannot be accomplished. Acquiring source code is a non-starter. This is the life blood of the software developer and with very few exceptions will not be, repeat not be, provided. HP will provide the source code associated with some of their devices but they also contain the clause that if you modify it and it causes problems with the devices the associated warranty becomes invalid.

Some laboratories use the software development cycle as a purchase medium and equate the successful purchase of an application package, through the process, as a verification/validation process. This is **not** V&V. How one purchases or acquires software has nothing to do with the V&V process. If the laboratory acquires software externally then the certificate provided with the package is adequate. The certificate states, "This Certificate of Authenticity is your assurance that the software that you have purchased with your computer is legally licensed from Microsoft Corporation. If you have any concerns about the legitimacy of this Certificate of Authenticity or the software your have received, call the Microsoft Piracy Hotline at 1-800-RULEGIT (in the U.S. or Canada) . . . " Other phrases contained on either licence certificates or contained in the provided documentation state: "We warrant that the storage media in the product will be free from defects in materials and workmanship...," "You may not modify the program...," and, "Translate, reverse engineer, decompile or dissemble the program...(and

Figure 2. Software Quality Factors *(6, 7)*

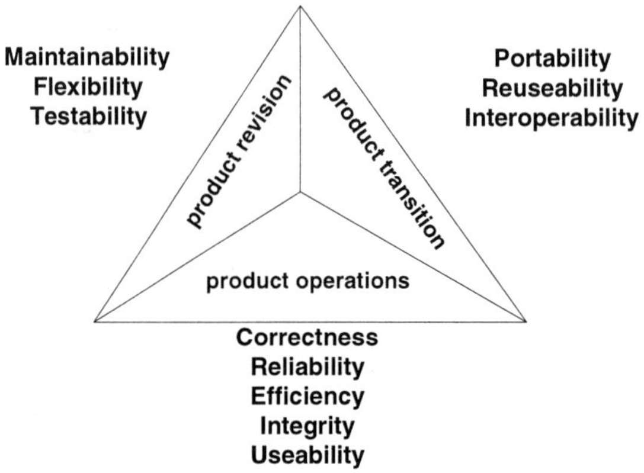

Maintability - Can I fix it?
Flexibility - Can I change it?
Testability - Can I test it?

Portability - Will I be able to use it on another machine?
Reusability - Will I be able to reuse some of the software?
Interoperability - Will I be able to interface it with another system?

Correctness - Does it do what I want?
Reliability - Does it do it accurately all of the time?
Efficiency - Will it run on my hardware as well as it can?
Integrity - Is it secure?
Useability - Can I run it?

this) is expressly prohibited by applicable law." According to these caveats even if the source code were to be provided nothing could be done with it. Recommend that in lieu of attempting to conduct an improper and invalid verification/validation process the user contact the various providers of software and ask for a verification certificate. The may or may not have one and they most likely refer to the licence certificate contained with the software as a valid verification certificate.

The one aspect of the V&V procedure that is independent of the commercially available software are those instances wherein the laboratory develops its own software. In this circumstance they are considered the developer and will be required to have for review and archive the complete set of documentation, that is, statement of need, source code, change logs, alpha and beta versions, final copy and future changes and procedure to effect version change. This satisfies the GLP requirement that any item used in the conduct of a study be available for inspection. If the software was developed for the express purpose of one study then all the material associated with that software will be archived with the study. If the software has use throughout the laboratory then the archive requirement will be for the entire package with a copy of the version used in a specific study to be archived with the study. The objective of retention is to permit the conduct of the study to obtain the same results of data submitted to EPA. The software and the associated documentation must be retained as per 40 CFR 160 §160.195, Retention of Records and 40 CFR 792, §792.195. Experience has shown that very few laboratories have the desire and capability to develop in-house software of any utility. Small macros for random number generation and selection are really not programs in the true sense but caution is suggested. If a laboratory uses such a macro, it is suggested that a version of it be included in the study archives. There is no requirement that it be included in the study report.

With respect to commercially produced software it is recommended that the original copy of the software and associated documentation be archived. This a fail safe suggestion. Within the United States there are a growing number of software escrow facilities. Some of the groups are club types where joining for a fee and yearly maintenance a certain type of software will be available into the foreseeable future. Clubs are organized along use line such as word processing, spread sheet, and graphic programs and within the set, specific products such as Corel Draw, WordPerfect, Word and Lotus. Major commercial escrow enterprises have the full range of popular software and use can be by yearly fee plus surcharge or by copy purchase. The escrow companies/clubs assure preservation of the most popular software. As time progresses they will possibly offer a service to convert old files created using an older software version to a current and useable version. This would permit record/files to be accessed at any time. The other alternative, for archiving study results, would be to convert all files to ASCII. The ASCII code is a standard code and does not change. ASCII will be in use for the foreseeable future and is a useable format for any record/file with or without the original software.

Software has been the main thrust of this presentation. The validation procedure for the hardware associated with the system is somewhat different. It is certainly hoped that no one other than trained technicians will attempt to validate the hardware using a physical method of validation. Hardware validation requires the use of sophisticated test equipment and tools. The attempt to do it by anyone not trained can cause the equipment to be severely damaged or destroyed. I have never seen at any facility inspected the types of equipment needed to do a systems validation. Nor have I met anyone in the GLP community capable of reading and/or reprogramming a chip. Leave the validation of the hardware to a trained technician.

The use of diagnostics is somewhat different. A majority, not all, of the computer manufacturers include a diagnostic disk with the purchase of the system. This can be used by almost anyone and is a great help when seeking either repair or guidance from a help desk. Most often a trained help desk technician can talk you through a corrective measure after you provide the output of the diagnostic run. The diagnostic disc can also be used as a method for obtaining a properly running device report. Please note that the term validation was not used. The diagnostic disc does not satisfy the validation procedure.

In summary, the terms verification and validation have a definite meaning in the computer science world and V&V should not be attempted by the uninitiated. Use the terms calibrate, audit or capable. Flow chart each device in the laboratory and its connectivity to other devices and systems. Most of the specific purpose devices have a built in test procedure to confirm that it is operating properly. The use of a dummy set of data with a known output can be passed through the system to determine proper functioning. The dummy data set would also indicate which device is not functioning properly and will narrow the problem considerably. Audit the system periodically to insure that it is operating properly and is being properly operated.

Therefore, with restatement of the theme of this Chapter - *Validation, Verification - Possibility, Probability*, are validation and verification needed, yes? Let the appropriate person or facility do it. Do not attempt to conduct a V&V procedure unless you possess the proper education, training and facilities. Is it possible for a laboratory to conduct a successful validation and verification process? I would have to agree that it is possible as all things are possible but I have to qualify the agreement by noting that the probability is very distant. As with my quotation from the Aeneid - do justice to what you know and do not attempt that which you do not. The deity of the computer has an abundance of antagonistic gods that will cause you never ending harm but with care and proper adoration, the benevolent gods always win out.

Literature Cited

1. *Handbook for Software Quality Assurance;* Schulmeyer, G. Gordon, CDP, and McManus, James I., Eds.; Van Nostrand Reinhold Company, New York, 1987.
2. *On the Development of Secure Software;* Lauer, H.C.; DCA/CCTC; 22 December 1976.
3. Programming Language Theory and Its Implementation; Prentice-Hall: Englewood Cliffs, New Jersey, 1988.
4. Beam, Walter R.; *Systems Engineering Architecture and Design;* McGraw-Hill Publishing Company: New York, 1990.
5. "Guidance on Software Package Selection;" NBS Special Publication 500-114; Frankel, S., Ed.; November 1986.

6. "A Framework for the Measurement of Software Quality," Proceedings of the Software Quality and Assurance Workshop, November 1978.
7. Cavano, Joseph and McCall, James A.; "Technology Assessment: Methods for Measuring the Level of Computer Security;" NBS Special Publication 500-133; October 1985.
8. Word, Excel, Access, PowerPoint, and Bookshelf are registered trademarks of Microsoft®.

Author Index

Subject Index

G

H

O

Q

T

T